T0202073

Communicating with Data

Communicating with Data

The Art of Writing
for Data Science

DEBORAH NOLAN
SARA STOUDT

OXFORD
UNIVERSITY PRESS

OXFORD
UNIVERSITY PRESS

Great Clarendon Street, Oxford, OX2 6DP,
United Kingdom

Oxford University Press is a department of the University of Oxford.
It furthers the University's objective of excellence in research, scholarship,
and education by publishing worldwide. Oxford is a registered trade mark of
Oxford University Press in the UK and in certain other countries

First Edition published in 2021

Impression: 1

Published in the United States of America by Oxford University Press
198 Madison Avenue, New York, NY 10016, United States of America

British Library Cataloguing in Publication Data
Data available

Library of Congress Control Number: 2020952658

ISBN 978-0-19-886274-1 (hbk.)
ISBN 978-0-19-886275-8 (pbk.)

DOI: 10.1093/oso/9780198862741.001.0001

Printed and bound by
CPI Group (UK) Ltd, Croydon, CR0 4YY

To Our Dads–Bill and Gary

Preface

C ommunication is critical yet often overlooked in the data-science pipeline. This book aims to help students and researchers write about their data insights in a way that is both compelling and faithful to the data. We address writing challenges specific to scientific investigations of data, such as how to describe data succinctly, create effective visualizations, write clean code, and accurately summarize statistical findings. We also provide more general advice on science writing, including how to distill findings into a story and organize and revise the story, and how to write clearly, concisely, and precisely.

Objective In our experience, university training in writing rarely addresses the challenges associated with technical writing. Our students lack opportunities to practice writing about their data-analytic processes and to learn from examples of good, domain-specific writing. To compound this problem, instructors of science courses typically have little experience in teaching technical writing. Many of us find it difficult to give students advice when we have only our personal experience with writing to draw from. In this book, we attempt to address both the teaching and learning challenges relevant to communicating the story behind a data analysis.

Audience We aim for this book to be a resource for students who want to learn how to write about scientific findings where the focus is on presenting the results of a data analysis. Instructors teaching a course in science communication can use it as a textbook, and others teaching a science course that has a writing component can use the book as a supplement. In addition, a researcher who is looking for help writing can use this book to self-train. Practicing statisticians, data scientists, or scientists who need assistance with writing about their data analysis findings will hopefully find guidance they can use to practice their communication skills in the context of their own work.

The only prerequisite is a knowledge of statistics at the introductory level. While we expect the reader to have at least a rudimentary understanding of statistics, the principles of communication found on these pages carry over to writing about more complex data analyses.

Examples and Activities Each chapter includes many examples and concludes with a collection of activities for practice writing. These examples and activities come from several scientific fields and a broad variety of publications. The main sources are scientific journals, but the advice is equally relevant to writing a report for a supervisor, a paper for an instructor, or an article for a popular magazine. To this point, we use the terms *article, paper,* and *report* interchangeably throughout. Additionally, many of the activities at the end of chapters give practice for those who want to write for a broader audience, such as a blog post or press release. We have honed the examples and activities in this book to focus on the essentials of writing about data and, at the same time, we have attempted to create scenarios that allow for individual creativity. Also, we use samples from our own writing and anonymized student work for examples of what not to do.

Organization The book consists of five parts. Part I aims to help the novice learn how to write by reading the work of others. We identify the main components of a data analysis, examine the argument, and point out how components of an analysis are organized into a story and written for a technical article. In addition, we read and examine material written for broader audiences, e.g., press releases and blog posts. Part II delves into the specifics of how to describe data at a level appropriate for publication, create informative and effective visualizations that support the main findings, and communicate an analysis pipeline through well-written, reproducible code. Part III demonstrates how to distill a data analysis into a compelling story and organize and write the first draft of a technical paper. Part IV addresses revision; this includes advice on writing about statistical findings in a clear and accurate way, general writing advice, and strategies for proof-reading and revising. Part V gives advice about communication strategies beyond the page, which includes giving talks, building a professional network, and participating in online communities. This part also provides over twenty portfolio assignments that are aimed at building upon the guidance and examples in the earlier parts of the book and continued writing practice.

We have provided online at http://communicating-with-data.github.io additional materials to use in a course. These include a detailed week-by-week syllabus that describes the topics covered, in-class activities, assignments, and additional reading. Along with the syllabus, we provide pointers to avoid potential problems with some classroom activities and ideas for grading written work. We also give ideas for how to use this book as a supplemental text in a science course or as the main text for a large course in technical writing where the student work is more limited in scope.

Acknowledgements This book grew out of a course on technical writing that we developed and first taught in the fall of 2017. Chapters 6 and 11 as well as portions of each of the Chapters 1–11 make up part of Sara Stoudt's PhD dissertation. We are grateful to the *Art of Writing Program* at UC Berkeley for providing a grant that enabled us to design and co-teach the course. We also owe a debt of gratitude to our students who willingly worked with us as we class-tested and developed the activities, assignments, and course materials. Their respect for each other, good humor, and diligence made the course a rewarding experience for all of us. Thanks also go to those who read earlier versions of this manuscript and tried it out with their students. In particular, we thank

Ben Baumer, Beth Chance, Kelly McConville, Amelia McNamara, Pamela Matheson, and Susan Wang.

The *Art of Writing Program* also supported our creation of this book by organizing a workshop where participants read and commented on all aspects of the book. We are much indebted to the workshop participants: Andrew Bray, Kathleen Donegan, Nicholas Horton, Ramona Naddaff, and Evan Variano, for their thoughtful and constructive comments that led to a much-improved manuscript.

Finally, thank you to all of the writers who agreed to let us use their work as examples throughout this book. We appreciate your willingness to have others learn from you and your writing.

Deborah Nolan and Sara Stoudt
July 2020

Contents

Part I
Reading to Write

Becoming a better science writer starts with reading, where we study how other scientists write. The idea is to start from the beginning and learn how to read and take notes, not simply as a consumer but as a creator–as a scientist writing about their data findings. We call this activity "reading to write".

The aim of Chapter 1 is to guide your reading of a technical article by helping you identify the main components of a data analysis, examine the argument the author is making, and recognize how these components are organized into a story. We examine how findings are motivated and how limitations are discussed, consider the level of detail provided about the data and findings, and read figures used to convey a core finding or justify an approach. We show how to pay attention to the details a writer uses to construct their argument and the choice of words they use in making claims. Knowing how to read as a writer will help you learn how to write well, correctly represent your findings, and generally become a more efficient writer.

While Chapter 1 focuses on reading formal science articles, Chapter 2 extends the reading-to-write process to other less formal genres, namely press releases and blog posts. If your main interest is in writing for a broader audience, we still recommend reading Chapter 1 first because it lays the groundwork for Chapter 2.

1 Reading Science Articles

L earning how to write effectively about data begins with a strategy for reading and examining how the author organizes and writes about their findings. When we read to write, we identify the main components of an analysis and notice how the author brings these components together to form a logical and compelling story. During this process, we discover examples and templates that we can use to organize our own work and write about our findings. To get started, the following three steps can be helpful.

- *Identify the elements of the data analysis.* We begin by looking throughout the article for various building blocks of the data analysis. We often find that some of the elements of a data analysis are included in the article, while others are not. This investigation helps us understand the choices that an author makes in writing about their findings, and, as a reader, we assess whether particular omissions impact the credibility of the conclusions or whether any included details are superfluous to the main story.

- *Examine the argument.* When we read an article, we expect to be convinced of the importance and validity of the findings. We look for context that explains how the findings fit with others' work, and we try to discern whether the new insights support, counter, or extend current views in the field. We consider the appropriateness of the analysis, the generalizability of the conclusions drawn, and whether others' work have been adequately and convincingly presented. To help make this assessment, we pay close attention to the words the writer uses and consider how an alternative, similar word choice could impact the strength of a claim. We also pay attention to the article's tone and whether it adds credibility and to the connections between paragraphs and between sections and whether they convey a cohesive story.

- *Map the organization of the document.* At a basic level, science articles have three main parts–a beginning, which serves to define and motivate the problem; a

Communicating with Data: The Art of Writing for Data Science. Deborah Nolan and Sara Stoudt, Oxford University Press (2021). © Deborah Nolan and Sara Stoudt.
DOI: 10.1093/oso/9780198862741.003.0001

middle that presents the findings and explains what they mean; and an end, which summarizes the conclusions and their importance. Mapping out these parts of the document helps us see how the author chooses to organize their analysis, gives a sense of what a reader might expect when reading an article, and provides templates that we might follow.

The remaining sections of this chapter describe how to carry out these steps for reading an article. We place ourselves in the position of a reader and identify what we look for when we read with an eye towards writing. Several examples are provided; some come from publications aimed at a broad audience, such as a popular statistics magazine and a journal focused on the history and culture of mathematics. Other examples are from more technical science journals but do not require a lot of background in the field to understand the core argument. Most of these articles are about four pages long so you can quickly read them in their entirety. Links to them can be found in the notes section at the end of the chapter.

This description of how to read an article also serves to introduce the goals of the book, and it provides a map of the topics covered in later chapters. Our main focus is on writing a technical article, but we also address how to write for other audiences. If general science communication interests you, then Chapter 2 extends the approach for reading technical articles to press releases and blog posts. Additionally, writing for these other "publics" is covered in designated sections and activities in Chapters 6–10.

1.1 Identify the Elements of a Data Analysis

If we are to be convinced of the conclusions of a data analysis, we must have enough information to assess the quality of the data. We need to know about the kind of data collected and the conditions under which they were collected. This information allows us, as readers, to evaluate the appropriateness of the methods used to analyze the data, decide how reasonable the conclusions that the author draws are, and assess the limitations, generalizability, and impact of the study. We look for the following aspects of the data and their analysis.

Design The question of how the data arose is the foundation to understanding an analysis. The data at hand can be generated from an experimental or observational process. Or, the analysis can be model-based and rely on summary statistics, or a meta-analysis that collectively examines several earlier studies.

Protocol The conditions under which the data were collected are critical. We want to know about the setting of the study, such as the time period, location, etc., and how subjects were selected, e.g., a probability sample, self-selection, or an administrative collection. Additionally, we also want to read about the outcome (i.e., response) measurements, covariates, and about nonresponse, including missing values, drop-out, and lost to follow-up. If appropriate, we also want information about treatments used, method of treatment assignment, and blindness of subjects and evaluators.

The description should be thorough enough to assure us that the analysis and conclusions are appropriate, and we expect the author to mention any limitations encountered in the data collection that might affect the conclusions. The amount of detail provided can vary a lot. For example, when the source is reliable and well-documented, such as a government survey, then we are satisfied if the author simply refers us to other, easily available documentation. Chapter 3 covers the topic of writing data descriptions.

Data Analysis There is an abundance of statistical methods available to analyze data, and our aim when reading to write is to observe how an author supports their findings through the presentation of specific results from the analysis. An analysis can be organized loosely into three categories: summary statistics and tables, statistical graphs, and formal methods, such as confidence intervals, hypothesis tests, and regression models.

A statistical graph is often used to convey the key finding of an analysis. Graphs are also provided as alternatives to tables of summary statistics and to support decisions made in choosing and assessing formal methods. Guidelines for making effective statistical graphs are provided in Chapter 4.

In addition to writing an article, the author often provides the code they wrote to clean and analyze the data. Writing code is another form of communication, and science writers are increasingly expected to make their code available for others to read, evaluate, and apply. A few simple guidelines for writing code to make it accessible and reproducible are provided in Chapter 5.

Complex statistical analyses often consist of a lot of assumption checking and modeling output, and choosing what information to provide and what form to provide it in is a crucial task for a science writer. For example, even when more advanced statistical methods are performed, summary statistics are often included to lay the groundwork for the analysis. Decisions about which technical material to include in an article is the topic of Chapter 6.

1.1.1 Determining if the Golden State Warriors Have Hot Hands

The *Mathematical Intelligencer* publishes articles written in an engaging, informal style aimed at what the journal calls a "general intellectual audience". For example in "Do the Golden State Warriors Have Hot Hands?" by Daks, Desai, and Goldberg, the researchers study the performance of star basketball players to assess evidence of a hot hand, i.e., whether a player is more likely to make a shot after having just made some shots. We consider how the authors describe the statistical elements of their argument for this broad audience.

At its core, the analysis consists of a hypothesis test. The authors gradually build to the formal definition of the test, starting with conceptual descriptions and examples. They first describe the test statistic in non-technical terms:

...the probability of making a shot given a string of two hits minus the probability of making a shot given an equally long string of misses.

Then to help the reader understand this definition, the authors compute the statistic for a concrete situation: Klay Thompson's shooting record in the December 23, 2016, game against the Detroit Pistons.

The record is represented by a string of 1's (hits) and 0's (misses):

$$1\ 1\ 1\ 0\ 1\ 0\ 0\ 1\ 1\ 0\ 0\ 0\ 0\ 0\ 1\ 1.$$

Thompson took 16 shots in this game, and as it happened, he made exactly half of them. ... There are four instances of two hits in a row, indicated by the string 11. We know what happened after the first three instances: Thompson hit the first time and missed the second and third times. But nothing happened after the fourth instance of 11 because the game ended before he could take another shot. ...we observed Thompson scoring one-third of the time following two hits in a row. In the other direction, given that Thompson missed twice in a row, he scored two-fifths of the time. ...The difference of the two conditional probabilities is 1/3 - 2/5 = -1/15.

After the conceptual definition and empirical calculation, the formal definition of the test statistic is provided and notation defined.

The underlying assumptions of the hypothesis test are essential; these assumptions lead to the probability model for generating data under the null hypothesis and calculating a p-value. The authors put forth a general explanation that the null distribution of the test statistic comes from "a random string of hits and misses". They further explain that the test is conditional on a given string of hits and misses for a specific player in a particular game, i.e., the null distribution is defined by all possible permutations of the observed string of hits and misses. The null distribution and p-value are estimated via simulation. A histogram of the null distribution is provided for Thompson's string of shots in the December 5, 2016, game against the Indiana Pacers. The non-normal shape displayed in the figure helps us understand the issues that arise with short sequences, which the authors aim to address with their methodology.

Additional statistical elements in the article consider alternatives to this hypothesis test. The authors calculate p-values with and without a strict inequality; drop free throws from the sequence of hits and misses to see if that changes the findings; use other models of randomness, such as generating hits and misses from a coin toss where the chance of a hit matches the player's record; and compute alternative test statistics. Few of the specific results from these variants of the hypothesis test are included in the article. Instead, the authors provide the reasons for considering the alternatives, non-technical descriptions of them, and a brief summary of the findings and how they support their conclusions. Given that the authors have explained the main test in detail, we have enough information to understand how they carried out the other versions of the test, and we can trust their findings.

1.2 Examine the Argument

A common misconception about science writing is that scientists do not make arguments when they write because science must be impartial and objective. However, the aim of a scientific article is to make a convincing case for the findings, and when we read an article, we are mindful of how the author makes their case. We become convinced of the importance of the work if we can judge its relevance and generalizability, see how the work fits with other results in the field, and determine that the author has written an honest and accurate portrayal of the findings. If we are to accept the findings as compelling evidence, we expect the scientist to accurately describe them, place them in the context of other findings, and demonstrate a balanced understanding of their implications. The art of organizing your findings into a convincing story is the subject of Chapters 6 and 7. Below are a few ways we examine the argument as readers.

Placement The findings from a data analysis need to be presented in context. In order to understand the relevance of new findings, we, as readers, need to know how these new results fit into existing theories. These findings may: support existing understanding and possibly fill a hole in current theories; address disagreements in the literature by providing an argument for one theory/approach over another; or advance a new explanation or a new method for understanding the world, which may express a critique of existing perspectives.

Generalizability Data are crucial to scientific arguments, and the quality of the data and the analysis are key to assessing the implications of new findings. An author should identify what they can say about their findings and what they cannot. Addressing both the limitations and the generalizability of findings go hand-in-hand, and we expect to see both in a well-written article.

Relevance The author must convince the reader that the identified problem of interest will help advance knowledge in a field and that the approach is relevant to the problem of interest. For findings to be considered impactful, they must reflect careful and sound data analysis, and the conclusions must be appropriate for the analysis— neither overblown nor too cautious. In short, the problem must be relevant to the reader, and the solution must be relevant to the problem itself.

Language A science writer must take care with the wording they use to make their argument. With the wrong choice of a word, an argument can easily slip into a stronger claim than what is supported by the data and analysis. When we read, we pay close attention to the precise wording of the author's statements, and reciprocally, the author must ensure that their writing accurately reflects the findings. Additionally, terms such as *significance* and *confidence* have special connotations in statistics so science writers should avoid them in their everyday usage. More information about careful use of statistical terms can be found in Chapter 8, and general advice for crafting sentences appears in Chapter 9.

1.2.1 Assessing a New Method to Weigh a Donkey

Case studies in *Significance* magazine offer examples of data analyses that are accessible to readers with differing levels of statistical training. These articles are generally engaging and relatively easy to read. They are written in a somewhat informal style, yet authors are careful in how they report findings and draw conclusions. As an example, we consider the article, "How to weigh a donkey in the Kenyan countryside" by Milner and Rougier. In this article, the authors describe a method they have developed to get a quick and accurate estimate of the weight of a donkey using only the measurements of the animal's length and girth.

We examine the wording in the article's two concluding paragraphs. The penultimate paragraph (below) presents the model and gives a concrete example of its application. We draw attention to three particular phrases: *Readers can confirm themselves*; *donkey number 78*; and *rounded to the nearest kilogram*.

We refitted the model with these recoded factors, taking the most populous levels of body condition and age as the reference (BCS = 3 and age ≥ 5). Our resulting model is: ...Readers can confirm themselves that a donkey of more than 5 years of age, with a BCS of 3, a girth of 122 cm and a length of 103 cm, weighs approximately 175 kg. This corresponds to donkey number 78 in our dataset, whose actual weight is 183 kg. The relative error is about +5%. The additive adjustments for those donkeys of different ages and body condition are given in Table 1, rounded to the nearest kilogram.

When the authors write *Readers can confirm themselves*, they informally invite us to work out an example to check our understanding of the method. To explain the assessment of the method in simple terms, the authors chose a precise set of input values that correspond to a donkey in the dataset, which they refer to as *number 78*. Of course, we don't really need to know the number of the donkey in the dataset. It is simply a device to engage the reader. To avoid confusion, the authors explain that the values in Table 2 are *rounded*.

The final paragraph below summarizes the accuracy of the new method using data that were set aside (not included in the model building exercise). Note the careful qualifications made by the authors with the use of the phrases *proceeds exactly*, *reasonable*, *typical*, *about*, *relatively*, and *over the range*.

Finally, we assessed our tool's accuracy, using the hold-out sample of one-fifth of the donkeys that we mentioned previously. The prediction of weight proceeds exactly as if we were in the field; that is, we used only the information in Figure 4 and Table 1 to obtain our predictions, which we could then compare to the donkey's actual weights. Figure 5 and Table 2 show that it is reasonable to claim that the typical accuracy of our tool is about ±10%, and that this is relatively consistent over the range of predicted weights from 75 to 200 kg.

The authors explain that the assessment is fair because it *proceeds exactly* as in the field. They summarize the accuracy by examining *typical* values; use rounded figures for simplicity (*about* ±10%); and evaluate the accuracy for donkeys similar to those for which they have data (*over the range*). Note also that their statement of results are softened by calling them *reasonable* and *relatively* consistent. Consider a sentence without these qualifiers: "We claim that the accuracy of our tool is 10%." Such a claim might leave the authors open to criticism when, e.g., the method is applied to a 50-kg donkey which is outside the observed range.

1.2.2 Helping the Scientific Community Adopt a New Standard

The article "Evaluation of the accuracy, consistency, and stability of measurements of the Planck constant used in the redefinition of the international system of units" by Possolo et al. helped usher in a new definition of the kilogram. This article was published in *Metrologia*, which according to the journal's website aims to "publish papers that contribute to the solution of difficult measurement problems and improve the accuracy of derived units and constants that are of fundamental importance to physics".

For the scientific community to redefine the kilogram, a consensus needed to be reached, and scientists needed to be convinced that the community was ready for the change. The Possolo study advocates for readiness to adopt the new definition. The investigators carry out a meta-analysis where they construct an argument from an evaluation of other studies. That is, their work does not propose a new method or present new data; instead, the primary contribution is a comprehensive and accurate portrayal of the work of others.

The introduction to the article lays out the main threads of their argument: assess-ment of the accuracy of the measurement of the Planck constant; consistency of the measurement across laboratories; stability of the measurement process over time; and a comparison of the reliability of the new kilogram measurement to the physical standard. The authors take these issues in turn and dedicate a section of the paper to address each.

The first avenue the authors consider is the question of stability. Measurements of the Planck constant have been made since 1979 at national laboratories across the world. These laboratories publish their measurements and their uncertainty, and the authors examine results from the various studies. A figure in the paper that displays a plot of the estimate (and its uncertainty) over time is particularly convincing for the redefinition because the measurements made over the past several years appear stable and accurate.

The accuracy and consistency results described in the article are situated on the world metrology stage. To lay the groundwork, these sections of the article each begin with a statement from the Consultative Committee for Mass of the International Committee Weights and Measures (CCM) on the prerequisites needed for adopting a new definition. For example, the accuracy section of the article quotes the CCM recommendation: "at least three independent experiments, including work from Kibble balance and XRCD experiments, yield values of the Planck constant with relative standard uncertainties not larger than 5×10^8." The authors argue that six experiments have achieved the CCM threshold.

The consistency section includes a critique of the commonly used approach to assess consistency. The authors point out the flaws of the "standard" statistic, consider an alternative, more-appropriate statistic from recent literature, and make the case that even with the flawed assessment the size of the inconsistency is negligible in comparison to accuracy.

Finally, the conclusions section of the paper makes several strong statements in favor of readiness for redefinition. This section first addresses reliability:

The level of agreement between the most accurate determinations of h available today is simply remarkable, especially considering that they involve multiple, different, and independent versions of two radically different approaches to its measurement.

The authors describe the accomplishment in glowing terms:

...a feat that has not been attempted since the national convention of the First Republic of France passed a decree on April 7, 1795, defining the gram as the 'absolute weight of a volume of pure water equal to the cube of one hundredth part of a meter, at the temperature of melting ice'.

The conclusions bring two additional points into the argument: there is a benefit to moving to this new definition with respect to the time and cost of dissemination; and continued research can improve the measurement techniques and further reduce uncertainty.

1.3 Map the Organization

A science article that reports on findings from a data analysis typically contains standard elements for presenting findings and making an argument, e.g., background information, methods description, summary of findings, and discussion of results. When reading to write, we note where in the article these elements appear.

When we read, we also look to the section headings to guide us and let us know what to expect and to reveal the organization of the story. When we notice placements of different types of findings, we can get ideas for how to organize our own story. Likewise, we want the wording of the title to accurately reflect the findings of the study.

Below is an overview of what to look for and where. More information on how to organize a technical article can be found in Chapter 7.

Context In an article on a statistical analysis, the author provides information about the data and the protocol used in collecting the data so that we, as readers, can use this information to assess the validity and importance of the findings. More specifically, we look for a description of the data (see Chapter 3) and how and why the data were collected. This description typically appears soon after the introduction to lay the groundwork for framing the analysis. Additionally, a literature review (see Section 7.5.1) places the current study in the context of related earlier findings. Review material often appears at the beginning or early in the middle part of the article. However, there are times when this information is embedded in a discussion (see Section 7.4.1) of the findings at the end of the article.

Methods The statistical methods presented in a report range from exploratory data analysis (EDA) to more formal analyses (see Section 7.3). EDA is typically discussed prior to presenting the formal analysis, and sometimes the findings from EDA are mentioned only briefly to confirm that the data are as expected. Other times EDA serves to motivate the approach taken in the formal analysis, and sometimes the article contains no formal analysis and consists entirely of EDA. A formal analysis might include an assessment of the effectiveness of a model, a comparison of findings to those that use a different, competing method, or a summary of estimator accuracy. These assessments and comparisons offer context that help quantify the impact and typically they come near the end of the presentation of the methods or in a separate discussion.

Statistical Graphs and Figures Visual displays can be useful in supporting study findings; they can justify a choice of method used or assess the impact of the analysis. Often a statistical graph, along with its caption (see Section 7.3.5), can make a compelling case for us to understand and judge the findings, and when such a central graph is placed early in the article it can help us readily grasp the main results without getting into the details of the analysis. See Chapter 4 for how to make and read graphs.

Impact Explanations of the impact of the analysis often appear in many places in an article. For example, the abstract (see Section 7.5.3) and introductory material (see Section 7.5.2) usually include a high-level description of the findings, the explanation of the analysis in the middle of the article often contains a more-detailed description of the impact of the findings and intermediate results, and the end of the article typically gives a high-level summary of the findings, their implications, and relationship to other work.

1.3.1 Organizing the Announcement of an Exoplanet

Astronomy & Astrophysics is an open-access journal that publishes articles in theoretical, observational, and instrumental areas of astronomy and astrophysics. In the article

"Transiting planet candidate from K2 with the longest period", Giles et al. describe new methodology developed to discover planets that have long orbital periods, and they apply these methods to data collected from the space telescope K2 to find a new exoplanet (a planet in another solar system).

The diagram in Figure 1.1 shows the layout and section headings of this four-page article. While this article is more technical than others in this chapter, we focus primarily on the paper's organization. The headings and their sequence in the paper follow general guidelines and reveal little about the specific content (introduction, observations, analysis, discussion, and conclusions). The abstract also includes standard identifiers: context, aims, methods, results, and conclusions.

The first paragraph of the introduction provides motivation for the new method and includes a brief literature review for context. A detailed description of the data appears in the observations section. Two figures are included in this section. The first of these figures provides compelling insight into the new methodology. This central graph shows an unusual dip in the light curve of a star, which the authors explain is how the transit of a star might be detected from K2 data. The second plot confirms that the structure observed in the first plot corresponds to a planet; this plot is based on follow-up data collected from the Euler telescope in Chile.

The diagram in Figure 1.1 shows that the formal modeling used to estimate stellar and planetary parameters occupies at least half the space in the article.

The discussion at the end of the paper focuses on the impact of the research. This section includes the caveat that it will take three years to confirm results. The relevance of the findings are conveyed in the claim that this is the longest orbital planet discovered to date, and in likening the planet to the gas giants in our solar system.

1.4 A Guide to Reading as a Writer

This chapter aims to prepare you to write by first helping you read from the perspective of a science writer. We chronicle how a science writer describes the analysis that forms the core of their scientific finding.

We don't expect you to read an article straight through from beginning to end. When we read science articles, we typically make several passes over the article, each time understanding the material in greater detail. One common approach is to begin by reading the abstract and introduction and then going on to examine the figures and tables and reading their captions. This first pass gives a high-level view of the results, and from there we might choose to read the discussion to understand the impact, the data description to assess whether the strength of the claims is justified, or the methods for more details on the proposed methodology. Eventually we read the entire article through from start to finish without jumping around. This complete read is especially useful to get a sense of how the author makes their argument.

In this section we provide guidelines to follow while reading that are parallel to the three aspects of reading a statistical report described in Sections 1.1, 1.1.1, and 1.3.

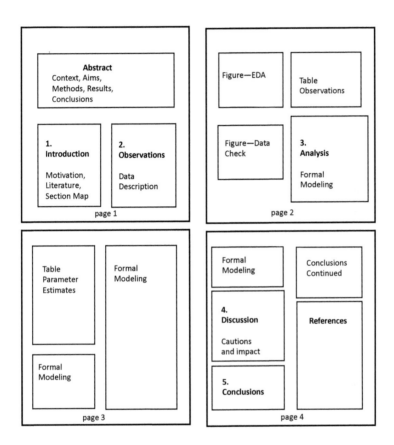

Figure 1.1 Layout of a Four-Page Article. This diagram shows the layout of an article on the discovery of an exoplanet. This organization is typical of a technical article published in a science journal.

1. *Statistical Elements.* A starting point for reading to write is to capture the analysis in a few broad strokes. The following tasks will help you in this process.

    a. *Summary.* Write a one- or two- sentence summary of the objective of the study.

    b. *Data.* Describe the data used in the study by following the guidelines provided in Section 1.1. If there are any aspects of the data collection and design not captured by these guidelines, describe them.

    c. *Analysis.* Briefly summarize the statistical analysis and the main outcomes. If possible, note any details excluded from the article.

    d. *Conclusions.* Write a short summary of the conclusions drawn from the data analysis.

2. *Argument.* Consider the following questions about how the writer positions their findings and their use of language.

    a. *Placement.* How do the results presented relate to the rest of the scientific literature? Do they support existing understanding of a phenomenon, fill in the gaps in current understanding, offer evidence in a debate, or provide a new approach or new evidence to study a phenomenon?

    b. *Generalization.* Are there limitations or difficulties in generalizing beyond the data? Are these limitations noted? Do you have any other concerns about the study?

    c. *Impact.* How does the study advance knowledge in the field? Do the author's claims seem adequately justified, overblown, or unduly cautious?

    d. *Language.* How do the specific words the author uses strengthen, clarify, and support their argument? Consider the following in answering this question.

        • Identify a paragraph that summarizes the result of the analysis. Examine the choice of words used, and relate them to the strength of the conclusions and the limitations the author places on the generalizability of the findings.

        • If you could change one sentence so that it is clearer, more concise, or more precise, which sentence would you choose? How would you change it, and what are your reasons?

        • How accurately and completely does the title of the article reflect the study findings?

3. *Organization.* Obtain a high-level picture of the organization by making a map of the article layout like the one in Figure 1.1. Mark the map with locations for the: objective and motivation; type of study and study protocol; description of methods and findings; statistical analysis; graphs and tables; related findings

made by others; discussion of limitations, implications, and generalizations; and conclusions. On the map, denote which pieces of the analysis are exploratory, formal, or a comparison/assessment.

Use the map to describe in a few sentences where the elements of the statistical analysis and the argument are found in the article. Specifically, address the following issues.

a. *Context.* What context is provided? Are related findings used to motivate the topic or are they provided in a later discussion?

b. *Methods.* How much space is dedicated to each type of analysis, i.e., exploratory, formal, comparison and assessment? Consider the level of detail in the method sections. What aspects of the analysis are mentioned without detail given, e.g., alternative models, assumption checking, confirmation of others work. How does the level of detail correlate to the background of the intended audience?

c. *Graphs and tables.* How are the graphs and figures used to support the findings? Do they address EDA, formal analyses, or comparison and assessment? Would additional figures be helpful? If so, what kind? Identify one key graph. Where is it located in the paper? What message does it convey with respect to the objective and conclusion of the study?

d. *Impact.* How often is the impact of the study discussed? Does the discussion of impact appear in the abstract, introduction, middle, and/or conclusions? Compare the level of detail in addressing the impact in these various sections of the article.

1.5 How to Weigh a Donkey in the Kenyan Countryside

As an example of how to use the guidelines in Section 1.4, we read and take notes on the article "How to weigh a donkey in the Kenyan countryside", by Milner and Rougier.

1. Identify the Statistical Elements

a. One-sentence summary. The aim of the study is to develop a simple tool to estimate the weight of a donkey from easily obtained measurements such as length and girth so that veterinarians in the field can prescribe medicine to sick animals without needing to weigh the animal directly.

b. Data. The data came from an observational study of donkeys. The study took place in July–August 2010 at 17 sites located in the Yaha and Naivash districts of rural Kenya. The participants were 544 donkeys from 2 species that were brought to the Donkey Sanctuary for de-worming. In some locations all donkeys were included in the study; in other locations a sample was taken. Pregnant donkeys and donkeys with visible disease were not entered into the study. Donkeys were measured by those working with the

Donkey Sanctuary. It is not stated whether there were multiple recorders. Recorders judged the condition of the donkey and scored them on a scale from 1 (emaciated) to 3 (healthy) to 5 (obese). The outcome is the weight (kg) of the donkey, and the covariates are girth (cm), height (cm), length (cm), sex (three categories: stallion, gelding, female), age (six categories), and body condition (potentially nine categories, but only six were observed). Three donkeys were excluded: one was too young, one was emaciated, and the third was obese. The results of the study were generalized to all donkeys in rural Kenya. Since no treatments were given, the question of blindness is not applicable. However, 31 donkeys were weighed twice to check the accuracy of the measurements. Those donkeys were unmarked after the first measurement so that the recorder wasn't aware of making a second measurement on a donkey. No information is provided about how donkeys were selected when too many arrived at the deworming station (e.g., whether a probability sample was used).

c. *Statistical Analysis.* The data were divided at random into a test set (one-fifth of the donkeys) and train set (the remaining four-fifths). The train set was used to select the model, and the test set was used to evaluate the model. The Box—Cox transformation was used to select a transformation for the outcome variable. The main statistical method was multivariate linear regression, which included categorical variables and used a modification of squared error to adjust for a different penalty for under or over estimating the weight. A small set of models were compared. The final product was a parallel-scale nomogram—a visual tool that converts combinations of two input values into an output value.

The article contains three figures and two tables. One figure shows a Box—Cox plot of loss to explain the choice of transformation for weight. A scatter plot of the size of the effect for the categorical variables justifies the collapse of some categories. The last figure contains a scatter plot of actual versus predicted weight for the test set to help visualize the relative error. The first of the two tables contains values for the size of the effect for the categories of body condition and age, and the second table gives the relative error in the test set.

In addition to the statistical graphs and tables, the article contains several other figures and diagrams, which add interest and help convey how the new method works. These visual displays are a photograph of two donkeys, a diagram of how to use a nomogram, a drawing of two loss functions for different medications, and a figure of the final nomogram constructed from the fitted model.

d. *Conclusions.* The model developed has a 10% error rate for donkeys with predicted weights in the 75–200 kg range.

2. Examine the Argument

a. *Placement.* The results are well situated in the scientific literature. The article includes a discussion about more complex nomograms and why they would be problematic in this case, and evidence that the nomogram method is in use for other kinds of animals. This study offers a new approach that maintains the simplicity of the two-variable nomogram, but offers more complexity by including additive effects.

b. *Generalization.* Generalizing to donkeys in rural Kenya seems reasonable. One caution: the donkeys in the study were brought to the Donkey Sanctuary to be dewormed, so the health of these donkeys might be different than the health of the typical donkey in rural Kenya.

c. *Impact.* The new tool advances the nomogram technique in a simple, novel, and useful way. The findings have the potential to greatly improve veterinary practice in rural areas.

d. *Language.* The title intrigues us by promising a solution to a logistical question. The second-to-last paragraph explains the results, and the final paragraph gives more information about assessing the accuracy of the approach. The wording in these two paragraphs is discussed in Section 1.2.1.

We correct the ambiguous language of the following two sentences: "This loss function is the cost of getting it wrong. It depends, among other things, on the drug that is being prescribed." The rewrite attempts to clarify what "it" means the two times it is used: "The cost of getting the dosage wrong is measured by the loss function. The cost depends, among other things, on the drug that is being prescribed."

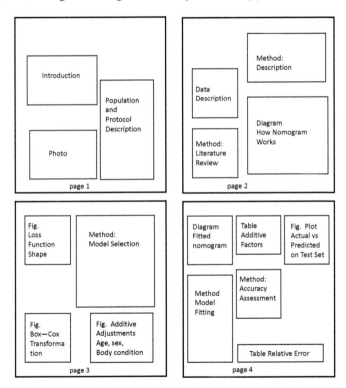

Figure 1.2 Layout of an Article on Weighing Donkeys. This diagram shows the article contains many visual components—figures, diagrams, and photo—to help convey the story to a broad audience.

3. Map the Organization The diagram in Figure 1.2 shows the layout of this four-page article. It locates the objective, study description, methods and analysis, figures and tables, literature review, discussion and conclusions.

a. Context. The topic is motivated in the introduction by a practical need of veterinarians in rural Africa. Additionally, after the description of the data, the notion of a nomogram and related work are provided. These references give motivation for the authors' novel statistical method.

b. Methods. More than half of the article is dedicated to the analysis, and about half of this space contains the formal analysis. The descriptions of all the methods are kept at a high level and little notation is used.

c. Graphs and tables. The core ideas, results, and assessment are explained with many diagrams and figures, which makes the material more broadly accessible. One key graph, a scatter plot of actual versus predicted weight for the test set appears at the end of the article and offers convincing evidence of the effectiveness of the method. A second key graph motivates the use of additive adjustments, which is the main insight of the authors' new method. This plot appears earlier in the article, where the method is described and the model fitted.

d. Impact. The impact of the study is discussed in three parts of the article. The discussion of impact first appears in the introduction as a what-if scenario, and then in the background literature where the new approach is contrasted to existing methods. Finally, the conclusion of the article discusses assessment.

1.6 Notes

The idea of learning statistics by reading news stories and science articles that report on statistical findings is described in Gelman and Nolan [5], and the guidelines in Section 1.4 are an adaption and extension of templates found there. Here the focus is on learning to write by reading technical articles from the perspective of a writer.

For more on how to examine arguments, see Graff and Birkenstein [9] and, in particular, the chapter called "The data suggest: writing in the sciences" by Gillen [7].

The articles used as examples in this chapter are openly available on the Web. More specifically, the article on hot hands in basketball by Daks, Desai, and Goldberg. [3] appears in the *Mathematical Intelligencer* (an earlier version [2] is available on arXiv). The article on the Planck constant by Possolo et al. [15] appears in *Metrologia*. Also, *The New York Times* provides a more general overview of the science and the people involved in changing the kilogram's definition (see Lim [12]). The article by Milner and Rougier [14] on how to weigh a donkey appears in *Significance*. The article by Giles et al. [6] on detecting planets orbiting other stars appears in *Astronomy & Astrophysics*. Additionally, a blog by Jorgenson [10] and press release [17] on the topic provide less technical descriptions about the new method for discovering exoplanets.

For more practice with reading to write, consider working on Portfolio Piece C.

1.7 References

[1] L. M. Brewer and J. J. Cochran. August's child is... favoured by fortune. *Significance*, 10(3):20–24, 2013. https://rss.onlinelibrary.wiley.com/doi/full/10.1111/j.1740-9713.2013.00661.x.

[2] A. Daks, N. Desai, and L. R. Goldberg. Do Steph Curry and Klay Thompson have hot hands? *arXiv*, 2017. https://arxiv.org/abs/1706.03442.

[3] A. Daks, N. Desai, and L. R. Goldberg. Do the Golden State Warriors have hot hands? *The Mathematical Intelligencer*, 40(4):1–5, 2018.

[4] I. Erotokritou-Mulligan, P. Sonksen, and R. Holt. Beyond reasonable doubt: catching the drug cheats at the London Olympics. *Significance*, 8(1):5–9, 2011. https://rss.onlinelibrary.wiley.com/doi/full/10.1111/j.1740-9713.2011.00469.x.

[5] A. Gelman and D. Nolan. *Teaching Statistics: A Bag of Tricks*. Oxford University Press, Oxford, 2nd edition, 2017.

[6] H. A. C. Giles, H. P. Osborn, S. Blanco-Cuaresma, C. Lovis, D. Bayliss, P. Eggenberger, A. C. Cameron, M. H. Kristiansen, O. Turner, F. Bouchy, and S. Udry. Transiting planet candidate from K2 with the longest period. *Astronomy & Astrophysics*, 615:L13,1–5, 2018. arXiv:1806.08757v1.

[7] C. Gillen. The data suggest: writing in the sciences. In G. Graff and C. Birkenstein, editors, *They Say / I Say*, pages 156–174. W. W. Norton & Company, New York, 2nd edition, 2010.

[8] D. Goulson. Neonicotinoids and bees: What's all the buzz? *Significance*, 10(3):6–11, 2013. https://rss.onlinelibrary.wiley.com/doi/full/10.1111/j.1740-9713.2013.00658.x.

[9] G. Graff and C. Birkenstein. *They Say / I Say: The Moves That Matter in Academic Writing*. W. W. Norton & Company, New York, 2009.

[10] A. Jorgenson. A new tactic for quickly finding exoplanets with Earth-like orbits. D-brief blog, July 2018. https://blogs.discovermagazine.com/d-brief/2018/07/20/how-to-find-earth-like-exoplanets/.

[11] J. LeSage, R. Kelley Pace, R. Campanella, N. Lam, and X. Liu. Do what the neighbors do: reopening businesses after Hurricane Katrina. *Significance*, 8(4):160–163, 2011. https://rss.onlinelibrary.wiley.com/doi/full/10.1111/j.1740-9713.2011.00520.x.

[12] X. Lim. The kilogram is dead. Long live the kilogram! *The New York Times*, November 2018. https://www.nytimes.com/2018/11/16/science/kilogram-physics-measurement.html.

[13] K. V. Mardia, F. Bookstein, and J. Kent. Alcohol, babies and the death penalty: Saving lives by analysing the shape of the brain. *Significance*, 10(3):12–16, 2013. https://rss.onlinelibrary.wiley.com/doi/full/10.1111/j.1740-9713.2013.00659.x.

[14] K. Milner and J. Rougier. How to weigh a donkey in the Kenyan countryside. *Significance*, 11(4):40–43, 2014.

[15] A. Possolo, S. Schlamminger, S. Stoudt, J. R. Pratt, and C. J. Williams. Evaluation of the accuracy, consistency, and stability of measurements of the Planck constant used in the redefinition of the international system of units. *Metrologia*, 55:29–37, 2018.

[16] S. Smith and M. Forster. The hazards of slavery. *Significance*, 8(4):179–182, 2011. https://rss.onlinelibrary.wiley.com/doi/full/10.1111/j.1740-9713.2011.00526.x.

[17] Universitè De Genève. Finding a planet with a 10 years orbit in a few months. EurekAlert!, American Association for the Advancement of Science, June 2018. https://www.eurekalert.org/pub_releases/2018-07/udg-fap071718.php.

1.8 Activities

1.8.1 Reading to Write

The case studies in *Significance* magazine offer examples of data analyses that are accessible to readers with a basic level of statistical training. These articles are generally engaging and relatively easy to read. One possible drawback with using them to study how to write is that they are written in a somewhat informal style. If this style of writing is different from the style of the journals and magazines for which you want to write, then you will probably want to select articles from other publications for this activity.

The following short articles are on a variety of topics—nature, economics, sports, public policy, and history. Choose one that interests you and write a summary that follows the guidelines in Section 1.4. Section 1.5 provides an example of how to do this.
"Neonicotinoids and bees: What's all the buzz?" (Goulson, D. [8]).
"Do what the neighbors do: Reopening businesses after Hurricane Katrina" (LeSage, J. et al. [11]).
"Beyond reasonable doubt: catching the drug cheats at the London Olympics" (Erotokritou-Mulligan, I. et al. [4]).
"August's child is... favoured by fortune" (Brewer, L. M.and Cochran, J. J. [1]).
"Alcohol, babies, and the death penalty" (Mardia, K. et al. [13]).
"The hazards of slavery" (Smith, S. and Forster, M. [16]).

1.8.2 Sharing Article Summaries

Write a one-paragraph synopsis of the article that you read for Activity 1.8.1 or for some other article of interest. This paragraph should encapsulate the goal of the analysis, the findings, and key features of the study. Ask someone who has not read this article to read your paragraph and discuss the study with you. (Or exchange summary paragraphs with someone who has read a different article and take turns discussing your articles.) Determine whether the reader understands the main points of the analysis, and identify what aspects they were unclear about.

As you discuss your synopsis, be sure to address the following questions.

- What did your reader misunderstand about the article from reading your synopsis? How could you change your summary to help them better understand?

- If you exchanged paragraphs, consider what you and your partner's articles have in common. How are they different? Consider the following questions.

  - Content. What types of graphs and statistical methods appear?

  - Detail. Does the author oversell or undersell the conclusions? How accessible do you find the description of the methods, considering the intended audience?

  - Organization. Where are the methods, conclusions, and discussion of potential flaws found?

Finally, rewrite your summary paragraph to address the inadequacies that were uncovered in your discussion.

1.8.3 Skipping Around as We Read

As mentioned in Section 1.4, at first we often do not read an article straight through from beginning to end. Instead, we make several passes over the article, each time understanding it in greater detail. In this exercise, you will get a sense of how much you can glean from a quick read of an article.

Choose an article from a journal in your field that has at least three statistical graphs. Read the abstract, introduction, and figures and their captions. Based on this limited knowledge of the paper, describe how each figure is used to help the authors make their argument. Select one figure that you think is core to the article. Explain your choice.

2 Reading Materials Written for Broader Publics

Communicating scientific findings to an audience beyond colleagues, collaborators, or researchers in our field plays an important role in supporting science. News stories offer one such venue for reaching a broader public, and press releases serve as valuable sources for these news stories. Press releases connect scientists to journalists. As the number of science writers in news organizations shrinks, a well-written and accurate press release can make it easier for a journalist to produce an accessible news story about an important technical topic.

Blogging has become a popular way for researchers to broaden their impact. A blog post gives a researcher an opportunity to communicate their experience performing a data analysis as well as the outcome of their analysis in a more informal style compared to technical articles. Bloggers explain by example, reveal their thought process, and offer transparency about their struggles. This additional narrative provides the context of an analysis and aims to help the reader understand the importance of the work. By focusing on intuition and motivation over derivations and proofs, a blog post can overcome the stigma that formal research may have among potential readers.

Since a blog post can be about smaller and light-hearted projects, it can provide insight into the behind-the-scenes decisions that a researcher makes but often does not reference in a formal manuscript. These decisions and the intuition driving the analysis process are often the hidden "art" of research that is learned via experience rather than via reading the literature or taking courses.

Reading examples of press releases and blog posts with an eye towards writing is a critical first step to writing for other publics, and we employ many of the strategies from Chapter 1. One key difference is that both press releases and blogs bring human interest aspects of research to the fore. Reading with a focus on language and the arc of the story can help us learn how to write engaging blog posts, and reading how expert opinions provide credibility, assess the impact of findings, and otherwise informally make an argument can help us learn how to write compelling press releases. Additionally, reading and comparing technical articles and related press releases/news stories can

Communicating with Data: The Art of Writing for Data Science. Deborah Nolan and Sara Stoudt, Oxford University Press (2021). © Deborah Nolan and Sara Stoudt.
DOI: 10.1093/oso/9780198862741.003.0002

help us figure out how to choose which technical elements to include or exclude from a press release and how to simplify descriptions of the statistical aspects of our work.

2.1 Reading Press Releases

As we read a press release, we consider the three aspects of reading described in Chapter 1: identify the statistical elements, examine the argument, and map the organization. These aspects take a somewhat different form for press releases because the audience is broader. Even though journalists read press releases, we effectively define our audience to be the readers of the journalists' news stories. For stories about data discoveries, we further narrow the audience to readers who are broadly interested in social, health, or physical sciences.

2.1.1 Identify the Elements–The Five Ws and H

A press release written in the USA typically begins with answers to five questions, referred to as the five Ws: who, when, where, what, and why? It can be confusing to answer these questions when the news story is about a data finding because some questions pertain to the analysis and others to the researcher. To help, we have expanded these five questions in Table 2.1 to specify whether the question refers to the investigator or the study. The what and why questions refer to the findings from the data analysis, but the other three are about the investigators. When a scientific study is the topic of the press release, then the who, when, and 'where' of the data are also relevant. These are described in Table 2.2. All together when we read a press release, we look for the answers to the combined set of eight Ws.

The answers to the eight W-questions often appear at the beginning of the press release in the introductory paragraphs. These answers can help us figure out how we might summarize our technical report for a general audience.

W	Question
Who	Who are the researchers investigating the problem and conducting the analysis?
When	When was the analysis carried out (e.g., "today", "this week", "recently", etc. are acceptable if the press release carries a date)?
Where	Where was the analysis carried out?
What	What were the findings from the analysis?
Why	Why are the findings important?

Table 2.1 The Five Ws. The traditional set of questions that are addressed in early paragraphs of a press release.

W	Question
Who	Who are the subjects of the study?
When	When were the data collected, e.g., when did the subjects participate in the study?
Where	Where were the data collected, e.g., where were the subjects under study located?

Table 2.2 Three Additional Ws. These questions about the data are also answered in press releases about scientific findings.

Example: *The Who, When and Where of a Prison Study*
In a press release about a survey of correctional officers in California prisons and jails, the first paragraph supplies the answers to all of the W-questions for both the investigators and the subjects of study.

California's correctional officers are regularly exposed to traumatic events that make them more likely to grapple with depression, PTSD and suicidal thoughts, according to a new report released this week by researchers at UC Berkeley's Goldman School of Public Policy. The first-of-its kind report is based on the 2017 California Correctional Officer Survey (CCOS) on Health and Wellness, a large-scale effort to gather individual-level information on the thoughts, attitudes and experiences of more than 8,000 criminal justice personnel, including prison guards and parole officers.

For the investigators we learn: who are the investigators–researchers; where are they–UC Berkeley's Goldman School of Public Policy; and when did they report their findings–this week (i.e., the week of August 23, 2018). For the subjects of the study, we read that the subjects are 8,000 criminal justice personnel, including prison guards and parole officers (who) in California (where), and that the survey was carried out in 2017 (when). Additionally, this introductory paragraph provides the what and why. What were the findings? California's correctional officers are regularly exposed to traumatic events and they have high incidence of depression, post-traumatic stress disorder (PTSD), and suicidal thoughts. Why are these findings important? This study is the first of its kind. ∎

The How Question The brief statements of what and why are usually expanded upon in the remainder of a press release and include additional details about the design, protocol, and methods (see Section 1.1). This information falls under the question: how? That is, how were the data collected and analyzed?

Example: *How an Experiment with Doctors and Patients was Carried Out*
A press release about a doctor's verbal reassurance speeding recovery time reports on the results of an experiment. In the middle of the press release, the essential elements of how the experiment was conducted are described.

To test the effects of a physician's words on patient symptoms, the researchers set up a simple experiment with 76 study participants.

They began their experiment by inducing a harmless, allergic reaction in study participants. A health care provider administered a histamine skin prick, a test traditionally used as a baseline to diagnose allergies. Histamine causes reactions like swelling, rashes and itching.

Following the skin prick, participants rated how itchy they were on a scale of zero to 100 at 3, 9, 12, 15 and 18 minutes after the skin prick.

Six minutes in, the health care provider came back in to check on the participants.

With approximately half of the participants, the provider assured them that "from this point forward your allergic reaction will start to diminish, and your rash and irritation will go away."

The details in these five short paragraphs are enough for us, as readers, to understand how the experiment was conducted. This information about the experimental protocol helps us assess the generalizability and reproducibility of the findings, which are described later in the press release. ∎

    In a press release there is no space for all of the information reported in the methods and results sections of a technical article, and it can be challenging to preserve the important aspects of the study. When we read a press release, we pay close attention to the few details provided. For example, the press release about the mental health of California correctional officers describes the type of study in its first paragraph. We read that the study was a survey, but no other details about how the survey was conducted appear in the press release. In this case, we might wish for a brief description of how the officers were selected, e.g., probability sample or census. As another example, large observational studies are often mined by many researchers. If the data were collected by someone else, we still expect the data source to be mentioned, but depending on the prominence and accessibility of information about the source, a simple statement that identifies the source may be adequate.

Example: *A Large Observational Study on Nurses*
An analysis of children with different religious upbringings and their health in their adolescence uses data from a large longitudinal study of nurses and their children. In a press release, the study is described as follows.

For this study, Chen and senior author Tyler VanderWeele, John L. Loeb and Frances Lehman Loeb Professor of Epidemiology, analyzed health data from mothers in the Nurses' Health Study II (NHSII) and their children in the Growing

Up Today Study (GUTS). The sample included more than 5,000 youth who were followed for between 8-14 years. The researchers controlled for many variables such as maternal health, socioeconomic status, and history of substance abuse or depressive symptoms, to try to isolate the effect of religious upbringing.

This large survey is described briefly because our common understanding is that a government study is scientifically sound. Plus, if we are interested, we know we can look up additional details in the study's publicly available documentation. ■

2.1.2 Examine the Argument

Quotations A press release usually contains quotations from the investigator and quotations from others who are knowledgable in the field and understand the potential impact of the findings. Quotations are typically interspersed with more detailed information about the statistical elements, and used to expand further on themes and introduce new information. Quotations also offer human-interest perspectives about the researchers and others involved.

When we read a press release or news story, we examine the quotations and how they help make the argument. Quotations from experts can convey a broader or different perspective than the investigator's, and they can serve to assess the impact of the findings. For example, others can say things about the importance of the findings that might be awkward for an investigator to say about their own work. In addition to adding credibility to the story, quotations can make the story more personable. When we read news stories, we generally like to hear directly from investigators through a quotation about why they work in their area and what they find exciting about their discovery.

Example: An Expert's Quotation on the Dearth of Research
The second paragraph of the press release about a study of the mental health of California correctional officers (see Section 2.1.1) includes a quotation from a 2013 Department of Justice report.

With few exceptions, the lives and well-being of correctional officers have been excluded from academic research and are absent from policy discussions about correctional programs and reform. As a comprehensive 2013 Department of Justice report aptly summarized: "Health and wellness among those who work in correctional agencies is an issue that has always existed, but is just starting to get the increasing attention that it deserves."

The quotation places the investigator's findings in the context of other work; that is, the quotation calls attention to the lack of knowledge in this area and the potential usefulness of this work. The Department of Justice is an external, credible source that validates the relevance of the study. ■

Example: *Quoting a Researcher's Hopes*

The two-paragraph introduction of the experiment measuring the impact of a doctor's verbal reassurance on recovery time (see Section 2.1.1) is immediately followed by quotations from the researcher that emphasize the importance of the findings.

"For many conditions, the simple act of being reassured by a medical professional can aid in the healing process, and we needn't always rely on medication and procedures to make us feel better," said Crum, whose research explores how patient mindsets can affect health outcomes and healing. "My hope is that findings like this one inspire additional research on the physiological mechanisms of assurance as well as promote training and compensation for physicians to more effectively leverage psychological forces in their practice."

The researcher provides a personal perspective about her hope for her research: an ambitious wish to see a major change in the way physicians are trained. The reader can easily relate to the researcher's aspirations to impact medical training, and this quotation makes a strong argument for the potential usefulness of the research. ■

Example: *Interspersing Quotations Among Study Details*

The third paragraph of a press release about student applications to charter schools breaks up the presentation of details with an indirect quotation from the researcher. This personal statement is mixed in with details about the Boston school district.

He says he was surprised that students with larger achievement benefits are less likely to apply to Boston's charter schools in the Boston Public Schools, Massachusetts' largest school district, which boasts an unusually large share of charter students.

Immediately following these details is another quotation from the researcher that explains why the differences are important.

"In the literature on school choice programs, it is often assumed that higher-benefit kids will be more likely to opt in to these programs, either because they may be less satisfied with the effectiveness of their other schooling options or because they have some information about whether the program is a good match," Walters says.

Here, Walters explains that his findings disagree with the assumptions in the prevailing literature about which children enroll in charter schools. One way that he makes a case for his findings is by admitting that he too was surprised by his findings. Notice also that he refers to the children in the study as "kids" which helps make his research seem less removed and more connected to the real-world problem. ■

Comparisons Press releases often include comparison numbers that help us put the findings in a context or scale with which we are familiar or that matters to us. This information conveys the relevance of the findings and, when provided by an expert, adds credibility.

Example: Concluding with the Larger Context
The final paragraph of a press release on the study of *Salmonella* in salad greens provides relevant information on the problem.

Research published recently by the Food Standards Agency reported that annually there are more than 500,000 cases of food poisoning in the UK. While poultry meat was the most common source of infection, some 48,000 of food poisoning cases were from fresh produce: vegetables, fruit, nuts and sprouting seeds. Importantly, *Salmonella* was the pathogen that caused the greatest number of hospital admissions–around 2,500 per year.

We read here how *Salmonella* is a large and potentially costly problem for people in the UK. These comparison numbers help us understand the problem in terms we can relate to, and they convince us of the importance of the research. ∎

Convincing Language Press releases tend to have minimal technical jargon and are written clearly, simply, and accurately. When a technical term is used, it is essential to the story and often defined through an analogy or example. This approach aims to simplify the story as much as possible and use strong terms to describe findings that do not exaggerate their importance. When reading, we pay particular attention to the author's substitution of technical jargon with simpler explanations. Getting accurate translations into everyday language can be one of the biggest challenges we face in writing press releases.

2.1.3 Map the Organization–The Inverted Pyramid

The W-questions (see Section 2.1.1) are often addressed in the first few paragraphs of a press release. However, the beginning of a press release also has the job of capturing our interest so that we continue reading. After we read about the essential elements of the researchers and their findings, we look for additional details about the research. These details are organized from most to least important in what is referred to as an inverted pyramid. Additionally, the details are interspersed with quotations that help make the argument for the findings and add a human interest component to the story (see Section 2.1.2). Finally, the press release finishes with a brief conclusion. In this section, we provide examples of this organizational structure of a press release.

The Hook A catchy opening is referred to as a hook. The first sentence of a press release should catch our attention and entice us to read more. While the first sentence needs to hook us, we are wary of exaggerated findings. Press releases sometimes leave it to the journalist writing the news story to provide the hook and simply begin with a straightforward presentation of the Ws.

Example: *Hooking the Reader with Why*
The first sentence of the press release about the effect of a doctor's words of reassurance on the speed of recovery from an allergic reaction begins with a single catchy sentence.

To feel better faster, a dose of reassurance might be just what the doctor ordered.

The reason(s) why the findings are important are conveyed through colloquialisms, *feel better faster*, and *just what the doctor ordered*–phrases that many readers are familiar with. We get the impression from this introductory sentence that offering verbal encouragement is an easy action to take to help patients feel better faster and could potentially have far reaching impact. This hook reveals that the results pertain to health, a popular topic; uses quirky language, e.g., *a dose of reassurance; and makes the relevance of the results easy for readers to grasp.* ∎

After the essential facts about the findings have been conveyed (i.e., the W-questions in Tables 2.1 and 2.2), the press release provides additional details, organized from most to least important. This inverted pyramid allows us to stop reading before the end of the article without missing the main point(s), and it allows the journalist to easily cut the press release and create a shorter story without losing the most important details. It can be a useful exercise to read a press release with an eye towards this kind of truncation. That is, we identify places in the story where a reader might stop, and we consider what would be their understanding of the scientific discovery at that point. In print journalism there are space constraints, so this cut often ensures that the essential elements are placed "above the fold" in the physical paper. In digital journalism, the above the fold analogy becomes the scroll (i.e., the essential elements must be readable before the first scroll).

Background Unlike in technical articles, background material typically appears towards the end of the press release. If the article is cut short, then the background material, rather than new findings, are lost. As we read, we note how the background material is conveyed, e.g., through quotations from the researchers and other experts in the field that place the work in context. Quotations can provide information in a way that is more personable and attention grabbing.

Example: *Adding Context About Boston Charter Schools*
Towards the end of the press release about the applicants to charter schools (see Section 2.1.2), one of the paragraphs provides relevant background information about why charter schools were created. While not a direct quotation, the researcher is credited with the insight.

One original motivation for charters was to allow more experimentation with new school models, which seems to have uncovered some very effective approaches and others that are less so, Walters says.

Another paragraph supplies details about how charter schools are run and their prevalence.

Most Boston charter schools follow a model that features extended instruction time, strict behavior codes, an emphasis on traditional reading and math skills, selective teacher hiring, and teacher monitoring—a combination that other research indicates boost student achievement. In 2010—2011, some 12 percent of Boston middle-schoolers attended charters, which then numbered 14.

This information brings context to the problem in terms of the goals of charter schools and the potential numbers of students affected. However, if needed, both paragraphs can be cut without losing the main point. ∎

Conclusion The conclusion of a press release often ends on a positive, forward-looking note. It is typically a short, one-or two-sentence paragraph that summarizes what has been said already and possibly includes a final dramatic quotation. When we read a concluding statement, we look to see whether it can be pulled from the end and placed earlier in the piece to create a shorter story.

Example: *Concluding with a Quotation*
The penultimate sentence of a press release on the study of *Salmonella* in salad greens offers a quotation from the chief executive for the agency that funded the research. In the quote below she explains why this kind of research is important.

Professor Melanie Welham, Chief Executive, BBSRC said: "Food-borne pathogens like *Salmonella* are serious bacterial threats that affect our health which is why BBSRC invests in research to understand and combat food poisoning."

Following this quotation is the paragraph (see Section 2.1.2) that provides comparison figures on the extent of the problem. Either the quotation or the final paragraph could be used to end a shorter news story. ∎

2.2 Reading a Press Release and Its Technical Article

A useful strategy when reading to write a press release is to also read the technical article on which the press release is based. Comparing these two pieces of writing can help us understand the choices the author makes in deciding what information to include and exclude from the press release. We also see which points from the technical article are conveyed through quotations, and how quotations support the argument or otherwise add to the story. We can compare the use of language between the article and press release to get a sense of how to report findings with minimal technical detail and less formal, but still accurate, language. Furthermore, a comparison of the organization of the press release and technical article gives further insight into how to write a science-based press release.

As an example, we read a short press release about the discovery of a planet outside our solar system (see Figure 2.1). The corresponding article was published in *Astronomy & Astrophysics* in June 2018, and the press release was posted the following month to the *Science News* site, EurekAlert!.

2.2.1 Identify the Elements

Many of the W-questions are addressed in the rather long subtitle:

Current techniques tend to only detect exoplanets with short orbital periods, however a new method developed by UNIGE researchers allows to find within months planets with periods lasting several years.

The subtitle describes the discovery made by researchers at the Universitè de Genève (the who and where for the investigators). The press release concerns the discovery of a new method for detecting exoplanets (what), and emphasizes its importance since current techniques are only able to detect exoplanets with short orbital periods (why). If we can assume that the date of the press release is close to the publication of the findings, then we also roughly know when the discovery was made.

Later in the article we read about the who and where of the data. The data are measurements from two telescopes: the space telescope, K2, and the Euler telescope in Chile. When the data were collected is not provided. The bulk of the statistical elements found in the technical article (how) do not appear in the press release. That is, the observations and analysis sections of the article describe in detail how the data were created from the telescope measurements, analyzed to estimate the orbital period, and validated. With the exception of a graph and some specific figures for a planet that was discovered, very few of these technical details appear in the press release.

2.2.2 Examine the Argument

We consider how the press release makes an argument in the absence of so many of the relevant statistical elements. The press release aims to convey the usefulness of a new method to detect planets with long orbits in other solar systems. The key feature of the innovation is that the method takes only a few months to detect a planet with orbits as long as 10 years, and this fact is mentioned several times in the press release.

The first two paragraphs compare the two methods and point out the problem with the existing technique, i.e., that a planet must be observed over three orbits to accurately detect it so the method cannot practically be used to detect planets with long orbital periods. The second paragraph makes the issue relatable by explaining that to detect a planet such as Jupiter the older approach would take at least 30 years.

The third and fourth paragraphs describe the new method. A quotation from Helen A. C. Giles, the lead researcher, explains a small part of the data collection process.

PUBLIC RELEASE: 18-JUL-2018
UNIVERSITÈ DE GENÈVE

Finding a planet with a 10 years orbit in a few months

Current techniques tend to only detect exoplanets with short orbital periods, however a new method developed by UNIGE researchers allows to find within months planets with periods lasting several years.

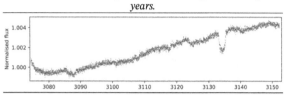

This is data from the light curve of the EPIC248847494 star.
The transit is clearly visible, on the upper right part of the image.

To discover and confirm the presence of a planet around stars other than the Sun, astronomers wait until it has completed three orbits. However, this very effective technique has its drawbacks since it cannot confirm the presence of planets at relatively long periods (it is ideally suited for periods of a few days to a few months). To overcome this obstacle, a team of astronomers under the direction of the University of Geneva (UNIGE) have developed a method that makes it possible to ensure the presence of a planet in a few months, even if it takes 10 years to circle its star: this new method is described for the first time in the journal *Astronomy & Astrophysics*.

The method of transits, consisting of detecting a dip in the luminosity of the host star at the time the planet passes, is a very effective technique to search for exoplanets. It makes it possible to estimate the radius of the planet, the inclination of the orbit and can be applied to a large number of stars at the same time. However, it has a significant limitation: since it is necessary to wait at least three passes in front of the star to confirm the existence of a planet, it is currently only suitable to detect planets with rather short orbital periods (typically from a few days to a few months). We would indeed have to wait more than 30 years to detect a planet similar to Jupiter which needs 11 years to make the full tour).

To overcome this obstacle, a team of astronomers led by researcher Helen Giles, from the Astronomy Department at the Faculty of Science of UNIGE and a member of the NCCR PlanetS, has developed an original method. By analysing data from the space telescope K2, one star showed a significant long-duration temporary decrease of luminosity, the signature of a possible transit, in other words, the passage of a planet in front of its star. "We had to analyse hundreds of light curves" explains the astronomer, to find one where such a transit was unequivocal.

Helen Giles consulted recent data from the Gaïa mission to determine the diameter of the star referenced as EPIC248847494 and its distance, 1500 light-years away from the planet Earth. With that knowledge and the fact that the transit lasted 53 hours, she found that the planet is located at 4.5 times the distance from the Sun to the Earth, and that consequently it takes about 10 years to orbit once. The key question left to answer was whether it was a planet and not a star. The Euler telescope of the UNIGE in Chile would provide the answer. By measuring the radial velocity of the star, which makes it possible to deduce the mass of the planet, she was able to show that the mass of the object is less than 13 times that of Jupiter—well below the minimum mass of a star (at least 80 times the mass of Jupiter).

"This technique could be used to hunt habitable, Earth-like planets around stars like the Sun" enthuses Helen Giles, "we have already found Earths around red dwarf stars whose stellar radiation may have consequences on life which are not exactly known". With her method it will no longer be necessary to wait many years to know whether the detected single transit is due to the presence of a planet. "In the future, we could even see if the planet has one or more moons, like our Jupiter," she concludes.

Figure 2.1 Press Release on Exoplanets.

The fourth paragraph provides a specific example of a planet that the researchers discovered. This planet has an orbital period similar to Jupiter's and is also similar to Jupiter in density. The parameters needed to calculate the orbit of the exoplanet are mentioned, such as the distance of the star from Earth and the travel time for the planet to pass in front of the star. This limited information is enough for the reader to grasp how the parameters are used to solve the problem, but specific methods are not provided. Additionally, a single graph depicts how the transit is revealed by the data. The article concludes with a final paragraph of quotations from Giles. The quotations aim to convey excitement about the potential for scientific discovery that this new method holds.

For comparison, the introduction of the original article provides context: there is a need for detecting exoplanets from a single transit to speed up discovery. This context is backed by references to other results and their limitations, i.e., existing approaches are practical only for exoplanets with short orbital periods. Many of the points made in the discussion and conclusions sections of the paper are conveyed in the press release through quotations from the lead researcher. For example, the comparison of the exoplanet to Jupiter and the possibility of being able to detect moons of the exoplanet appear as quotations in the last paragraph of the press release. The technical article conveys the potentially large number of exoplanets that could be discovered by this method and explains that it will take three years to confirm their findings for the Jupiter-like exoplanet. Neither of these points appear in the press release. The former may be unnecessary for the general audience, but the latter seems newsworthy. The technical article also mentions the possibility of detecting habitable planets and moons and references the technical requirements needed to accomplish this goal. This information is also conveyed in the press release via an enthusiastic quotation from Giles.

2.2.3 Map the Organization

We studied the organization of the associated technical article in an earlier example (see Section 1.3.1). The article has a traditional organization, which makes it easy for a reader to find the expected pieces of information. The press release consists of a title, subtitle, one graph, and five paragraphs. As mentioned already, four of the Ws are addressed in the detailed subtitle. The graph serves to reveal the structure in the data that corresponds to the exoplanet crossing in front of its star.

An unusual feature of the organization of this press release is that the first two paragraphs contain a considerable amount of background information about the existing method, and there is quite a lot of overlap between these paragraphs.

2.3 Reading Blogs

Unlike with technical articles where the audience is defined by the choice of journal, a blogger has more control over who they want to reach. Bloggers can essentially define their audience and tailor their writing style to fit the expectations of the people they are trying to reach. There are many reasons for writing a blog post, and the post format varies widely depending on its purpose. A first task when we read a blog post is to identify the motivation of the author. Then, we can focus on reading those posts that match our reason for writing.

Given the nature of online communication, a blogger has greater pressure to engage and entertain their readers. When we read a technical report, we often skip around and only read the technical details we are interested in. Blog reading is different–typically we read sequentially (from start to finish) and only make it to the end of a post if we are consistently engaged. Otherwise, we click away to another page. When we read a post, we look for the arc of the story and other strategies that the author uses to keep our attention and keep us scrolling, i.e., we examine the argument.

2.3.1 Identify the Motivation

There are many reasons for writing a blog post. A writer may want to (1) make their work more accessible, (2) teach a concept, (3) help others avoid having to reinvent the wheel when solving a technical challenge, (4) foster discussion, (5) synthesize an experience, or (6) represent themselves in their field. Identifying the author's motivation helps guide our reading of their blog post.

Blog posts should be accessible, yet rigorous, and bring an element of personal excitement to a topic. For example, if the author's aim is to make their work more accessible than the technical report they have written, then the results should not be discussed in a dry and removed manner. Nor should the post simply be a compressed version of the technical report. Think back to the hook from Section 2.1.3; the blog post should have a creative angle beyond the purpose of summary.

Certain aspects of a research-level analysis may be beyond the scope of a blog post, but we still need to understand what pieces of a sound research study have been left out for expository convenience. For example, a post may introduce a new methodology and to make fair comparisons, the data should be normalized, but the normalization complicates the analysis. The blog post can focus on the simple process, avoid drawing conclusions from the limited analysis, and briefly acknowledge how to properly handle the data. Even though the content in a blog is often simpler and less serious than formal technical work, the limitations of the research still need to be communicated.

2.3.2 Examine the Argument

Each of the reasons to write a blog post mentioned in Section 2.3.1 leads to particular writing challenges. Bloggers use a variety of strategies to overcome these challenges and make a persuasive argument. We point out a few to take note of as you read.

Make Work Accessible When a scientist wants to make their work accessible to a reader outside their field (or new to the field), they need to convince us, the readers, that it is worth investing time in learning about the topic. This type of post is most similar to a press release, valuing parsimony in the choice of details, relying on a strong narrative, and avoiding jargon. Although blog posts have a narrower focus, we still consider whether the writer has included the most important elements that make up the storyline. We also consider whether the problem is put in context and whether the writer makes clear who the work affects and how. When reading this type of blog post, we keep an eye out for technical terms and assess whether the writer finds a way to explain them, at least at a high level, simply.

Example: *Using Narrative to Explain Multi-Criteria Decision Analysis*
The *Methods in Ecology and Evolution* journal has an accompanying blog where researchers write a blog post as a teaser for their formal paper. The excerpt below explains the fundamental problem in context (a need to compare different things that can't be judged by the same criteria) and identifies a tool to solve the problem (multi-criteria decision analysis).

In real-life situations, it is far more common for decisions to be based on a comparison between things that can't be judged on the same standards. Whether you're choosing a dish or a house or an area to prioritise for conservation you need to weigh up completely different things like cost, size, feasibility, acceptability, and desirability.

Those three examples of decisions differ in terms of complexity—you'd need specific expert knowledge and/or the involvement of other key stakeholders to choose conservation prioritisation areas, but probably not to pick a dish. The bottom line is they all require evaluating different alternatives to achieve the desired goal. This is the essence of multi-criteria decision analysis (MCDA).

Here we get context for the problem that the author's new method is trying to solve. The writer connects a problem in their field (how to choose areas of interest when weighing multiple factors) with decisions that the average reader makes every day. The following sentence explains the formal method simply, avoiding over-complicated jargon.

In MCDA the pros and cons of different alternatives are assessed against a number of diverse, yet clearly defined, criteria.

The reader now knows enough about the method to understand the rest of the post. ∎

Teach a Concept When a post aims to teach a concept, clarity can be challenging because readers may have different background knowledge. A good teaching blog motivates learning, uses specific examples and applications, and contains empathy. When we read this type of blog post we assess whether the author tells us why we invest time in learning about the topic and how the post will benefit us. We look for an explanation grounded in an example that moves from the abstract to the concrete or vice versa and helps us conceptualize the new information. We also note the absence of jargon and judgment words such as "clearly" and "it is obvious". For an example of this type of post see Section 8.10.8.

3. Avoid Reinventing the Wheel When a scientist spends a lot of time figuring out how to do something (particularly something having to do with software), they may want to contribute to the scientific community by writing a blog post that assists others to more readily accomplish that task. A successful post that helps others avoid having to reinvent the wheel gives context to the steps involved including enough detail, explanations, and identification of potential pitfalls. As we read, we make sure that we can follow the author's steps, obtain the same result or conclusion, and run the code provided without additional code or files not available in the post. We look for commentary that helps us not only understand the overall outcome but also follow the thought process throughout. Additionally, we note the author's transparency about what issues they ran into and how they solved them.

Example: *Providing a Comparison of Implementations of Logistic Regression*
In a blog post, the author explains why she made a comparison of various libraries that implement the same statistical method.

Say I want to use some statistical method. I have a few options. I could code it up from scratch myself, knowing that this might have undetected bugs and be pretty slow. I could Google what I'm looking for and use the first thing I find; similarly, there are no guarantees. Or, I could do my research, find all the packages that seem to offer what I'm looking for, and decide which looks best based on how thoroughly they've documented and tested their code.

But I ran up against a problem: there are multiple ways to do logistic regression in Python that seem equally good. I dig into a more in-depth comparison of the methods here.

This part of the blog post provides an explanation of the exercise in general. We get to see the thought process of the author. Simpler ways to choose an implementation were unsuitable because there was no obvious winner, so she had to do a more extensive comparison.

After the comparison, the author presents us with her final choice (see below). This information allows us to skip having to make a similar comparison for our own work.

For my purposes, it looks like the statsmodels discrete choice model logit is the way to go. It's significantly faster than the GLM method, presumably because it's using an optimizer directly rather than iteratively reweighted least squares. I'm not doing prediction and am not worried about overfitting, so the fact that I can't disable the regularization in scikit-learn is undesirable. However, it's the fastest of the three methods.

Here we are given an answer to the initial question but with a careful caveat about a potential pitfall. The author provides an explanation of her decision-making process (for her purposes she chose a faster speed despite the undesirable regularization), and we have been given enough detail to weigh whether this decision is also important. ∎

Spark a Discussion When an important debate arises in a scientific field, a researcher may want to contribute to the debate via a blog post that could gain widespread attention within and outside of their field. The challenge for the blogger is to clearly and succinctly explain a wide topic and then take a stance. This type of post should orient the reader, ask a specific question, and signal a point of view. A successful discussion blog post outlines both (possibly, multiple) sides, poses a specific question that narrows the focus from a potentially wide topic to the heart of the matter, and provides a clear and straightforward explanation. For an example of this type of post, see posts by Gelman.

Synthesize Understanding At times, a scientist wants to share a newly acquired understanding of a topic. This understanding often arises after consulting a variety of materials including papers and conversations with others. Science bloggers can contribute to the scientific community by bringing together ideas from different places. This is analogous to the avoiding reinventing-the-wheel motivation but for concepts rather than technical details. However, it can be a challenge to go beyond making a basic summary of the material on the subject. A successful synthesis type of blog post identifies the major take-aways and remaining gaps and makes connections between information the author obtained from different sources and relevant information from other contexts. When we read these posts, we look for signals of what information was most influential to the author and what information they wish had been included in the materials they are talking about.

Bring Perspective A scientist's professional identity can involve more than their scientific training, and another way they can contribute to the scientific community is to describe their challenges and successes for others to learn from and relate to. The main challenge in writing a blog post for representation purposes is to maintain professionalism but allow one's personal story to shine. When we read this type of blog post, we look for shared experience and honesty. Does the author describe their path to their current field or career? What were the turning points as they moved forward? What support networks helped them? The most powerful posts of this kind are written by authors who avoid sugar-coating their path if it was nontraditional or nonlinear. The author is not afraid to talk about the lows as well as the highs, reassuring readers who may be considering a similar path. Section 11.4 provides example posts of this type.

2.3.3 Map the Organization

Although a blog post is typically much shorter (300–600 words) than a formal piece of writing, the structure is similar. The introduction and conclusion give a preview and summary respectively, and the middle contains the details.

Blog posts vary widely depending on the motivation for writing and the style of the writer, but many have the following components.

- *Hook*: A hook is the way into the story. It can be a provocative statement or question, an unexpected phrase, or an opening anecdote. The goal is to get us intrigued enough to keep reading.

- *Angle*: Since a blog post is short, there has to be a narrower focus. An angle is a specific approach to a story; it can also help a reader identify the post's motivation.

- *Content*: The middle of a blog post contains the main content but stays succinct. The content might build up to a final, shining example or be a series of shorter examples that are tied by a narrative.

- *Wrap up*: Towards the end of a blog post there is often a recap that covers the important take-aways from the post. This serves as a reminder to the reader of what we just read in case we skimmed a bit too fast.

- *Teaser and resources*: The very end of a blog post often contains links to related posts by the author or others, a teaser for future posts, or other resources for an invested reader in case we want to follow-up on details left out of the main post.

2.4 Reading a Sample Blog

In this section we inspect the blog post in Figure 2.2 for the features discussed in the previous section.

2.4.1 Identify the Motivation

The motivation of this blog post is to help readers learn from the writer's mistakes by telling a candid anecdote of a stumbling point. Although the author discusses her work at a high level, the work itself is not the focus of the blog post. The author's goal is not to make the work itself accessible, but rather to make the workflow accessible. The post primarily brings a personal perspective to the author's work habits.

2.4.2 Examine the Argument

The use of "[high level space 1]" and the replacement of technical elements with capital letters are part of a strategy to abstract away technical details that the writer doesn't want to focus on.

"Does it go from [high level space 1] to [high level space 2] or from [high level space 2] to [high level space 1]?" he interrupts, like he had a better idea of what my program does than I do.

Understanding Through Documenting: The importance of seeing how your code fits into the bigger picture and how documentation can help facilitate that.

Last month, I showed up to badminton practice hoping to have some fun, get some exercise, and make some friends. I instead made a fool of myself, but learned a lot from the experience.

I met a computer science major (highly likely, the badminton team is overwhelmingly CS), working in the same lab as me (highly unlikely). Turns out, he was the student my advisor had in mind when he asked me to finalize my code to serve as an example. So we talked about my code. The conversation went something like this:

Him: "I didn't understand it."

Now he had my full attention. I thought the code was clean and well-modularized with descriptive function names and even some comments. How hard did he try to understand it?

Him: "What does it do?"

At this point, I realized I couldn't answer his simple question. I had no idea what this program, as a standalone entity, did. In that case, I could not have conveyed its high level purpose in the code or documentation, and he certainly was right to be confused.

I went into finals exam mode, trying to recall everything about the program.

Me: "Okay, so it turns A into B so that we can use B in C..." I paused.

I tried my best to quickly repicture the entire script and started piecing together what the functions achieve together.

Me: "We take this from A and call X on it to get A-prime, and then we call Y..."

Him: "Does it go from [high level space 1] to [high level space 2] or from [high level space 2] to [high level space 1]?" he interrupts, like he had a better idea of what my program does than I do.

Still unable to answer, I kept going through my memory of the code. Eventually, after a lot of starts and stops,

Me: "Yes, it goes from [high level space 1] to [high level space 2]." He was happy with my answer, but I was well-aware of how close I was to failing to explain my work.

I knew that I could get caught up in accomplishing my tasks and struggle with understanding where my work fits into the bigger picture, but this was the call to action I needed to prioritize this problem. In research, getting tasks done is only part of the bare minimum – with so many research directions, how can anyone trust my research if I don't have a clear idea of why I am doing what I am doing? The open nature of research means there is no clear path, so each decision needs to be well-justified.

That week, I took time to understand what exactly I was doing and why I was working on each task. I ramped up my documentation for that code to include a short blurb addressing what it does and what it is useful for, in addition to how to run the program. I realized that documenting my code in this way makes me think about the purpose of the code, and enables users to better and more quickly understand how the individual lines of code achieve the ultimate goal. I hope that by continuing to improve my documentation habits, I will become a better researcher and avoid embarrassing encounters like this in the future.

Figure 2.2 Blog Post on Code Documentation. This blog uses dialog between two badminton players to tell a story about the importance of good documentation.

Getting into the weeds would distract us from the overall message. The writer has revealed the "characters" of the story without getting into their full back story unnecessarily.

The dialogue between the blogger and fellow computer science major is a special case of using narrative. The writer inserts us directly into the conversation and provides commentary to add further context to the scene.

At this point, I realized I couldn't answer his simple question. I had no idea what this program, as a standalone entity, did.

We see honesty in this blog post's recount. As readers, we can empathize with this low point while reflecting on our own habits.

2.4.3 Map the Organization

The author starts the blog post by setting the scene. The hook is that she went to badminton practice ready for some exercise but ended up with a lesson for her professional life.

Last month, I showed up to badminton practice hoping to have some fun, get some exercise, and make some friends. I instead made a fool of myself, but learned a lot from the experience.

The blog post then continues as a dialogue between the author and a colleague with commentary interjected occasionally. The commentary helps us identify the angle: showing the impact of being confronted about work by someone who doesn't understand it.

I knew that I could get caught up in accomplishing my tasks and struggle with understanding where my work fits into the bigger picture, but this was the call to action I needed to prioritize this problem.

The blog post closes with a summary of lessons learned and an overall moral.

I realized that documenting my code in this way makes me think about the purpose of the code, and enables users to better and more quickly understand how the individual lines of code achieve the ultimate goal. I hope that by continuing to improve my documentation habits, I will become a better researcher and avoid embarrassing encounters like this in the future.

2.5 Notes

Our focus in Sections 2.1 and 2.2 is on reading to write press releases rather than news articles because we consider our audience to be scientists who want to write about their own findings.

There are many resources online for how to write a press release; we particularly liked the presentation by the University of Leicester [20]. We also benefited from conversations with Christoph Drösser who was the journalist in residence at the Simons Institute at UC Berkeley in 2017.

The inverted pyramid and the five Ws (and H) are popular constructs for describing the structure of press releases and news stories. We extended these by an additional three Ws (see Table 2.2) in order to address the special situation of a press release about a scientific study. The inverted pyramid is not the only structure for a news story/press release. See Section 11.7.1 for examples of profiles, where the person, not a news event, is the story and so the structure is different from the inverted pyramid.

The press releases used in examples and activities in this chapter were all obtained from websites of various universities' news offices. These include: press releases on charter schools [13] and correctional officers [18], which can be found on the UC Berkeley news site; the press release about a doctor's reassurance [5] that appears on the Stanford News Service; the story about religious upbringing [17] at the Harvard School of Public Health media site; and the press release on *Salmonella* found on the University of Leicester press office site [21]. The press release on exoplanets appears on both the Univerisitè de Genève media relations site and the American Association for the Advancement of Science (AAAS) site EurekAlert! [19].

For more information about scientific blogging, see Wilcox, Brookshire, and Goldman [22]. Example blogs abound on the topic of how to write a blog; these include Onibalusi [14], Klems [11], and Careathers [3], and those geared to academic writing, such as Dunleavy [6] and Cayley [4].

The blog posts used as examples in this chapter include some written by graduate and undergraduate students (Ottoboni [15], Jann [10]), as well as the professional post by Esmail and Geneletti [7] that appeared on the *Methods in Ecology and Evolution* blog and is based on their paper [8]. The blog in Figure 2.2 was written by a student in our Blogging for Data Science course (Jann [10]).

Gelman [9] provides many examples of posts aimed at sparking discussion. For an example of a blog post that synthesizes understanding see Peng's post [16].

The excerpts of blog posts used in Section 2.7.4 are by Averick [2], Andreas [1], and Mackintosh [12].

For more practice, consider working on Portfolio Pieces L, P, or Q.

2.6 References

[1] J. Andreas. Learning to reason with neural module networks. http://bair.berkeley.edu/blog/2017/06/20/learning-to-reason-with-neural-module-networks/, June 2017.

[2] M. Averick. NBA Horserace, April 2018. https://maraaverick.rbind.io/2018/04/nba-horserace/.

[3] L. L. Careathers. How to write a blog post-the ultimate guide. https://smartblogger.com/how-to-write-a-blog-post/, 2019.

[4] R. Cayley. Explorations of style. https://explorationsofstyle.com/.

[5] M. De Witte. A doctor's reassurance speeds healing from an allergic reaction, find Stanford psychologists. Stanford News Service, September 2018. https://news.stanford.edu/press-releases/2018/09/04/doctors-reassurance-speeds-healing/.

[6] P. Dunleavy. Writing for research. https://medium.com/@write4research.

[7] B. A. Esmail and D. Geneletti. Applications of multi-criteria decision analysis in conservation research. https://methodsblog.com/2018/03/23/mcda/#more-6437, March 2018.

[8] B. A. Esmail and D. Geneletti. Multi-criteria decision analysis for nature conservation: a review of 20 years of applications. *Methods in Ecology and Evolution*, 9(1):42–53, 2018.

[9] A. Gelman. Bit by bit: Social research in the digital age. http://andrewgelman.com/2018/04/14/36778/, April 2018.

[10] T. Jann. Understanding through documenting: The importance of seeing how your code fits into the bigger picture and how documentation can help facilitate that. https://stat198-spring18.github.io/blog/2018/04/03/understanding-through-documenting, 2018.

[11] B. A. Klems. The 12 dos and don'ts of writing a blog. https://www.writersdigest.com/online-editor/the-12-dos-and-donts-of-writing-a-blog, 2012.

[12] J. MacKintosh. Data driven DIY: hardwired...to shelf construct. https://www.johnmackintosh.com/2018-01-20-data-driven-diy/, January 2018.

[13] K. Maclay. Study asks why students with more to gain from charter schools are less likely to apply. Media Relations, UC Berkeley, April 2018. http://news.berkeley.edu/2018/04/03/study-asks-why-students-with-more-to-gain-from-charter-schools-are-less-likely-to-apply/.

[14] S. Onibalusi. 7 tips for writing that great blog post, every time. https://www.huffingtonpost.com/william-morrow/7-tips-for-writing-that-g_b_10724558.html, 2016.

[15] K. Ottoboni. Which logistic regression method in Python should i use? http://www.kellieottoboni.com/posts/2017/07/logistic-regression-python/, July 2017.

[16] R. D. Peng. Thoughts on David Donoho's "Fifty Years of Data Science". https://simplystatistics.org/2017/12/20/thoughts-on-david-donoho-s-fifty-years-of-data-science/, December 2017.

[17] C. Sweeny. Religious upbringing linked to better health and well-being during early adulthood. Harvard T. H. Chan School of Public Health, September 2018. https://www.hsph.harvard.edu/news/press-releases/religious-upbringing-adult-health/.

[18] UCB Public Affairs. Correctional officers at high risk for depression, PTSD, suicide, survey finds. Media Relations, UC Berkeley, April 2018. http://news.berkeley.edu/2018/08/23/california-correctional-officers-at-high-risk-for-depression-ptsd-and-suicide-new-survey-finds/.

[19] Universitè De Genève. Finding a planet with a 10 years orbit in a few months. EurekAlert!, American Association for the Advancement of Science, June 2018. https://www.eurekalert.org/pub_releases/2018-07/udg-fap071718.php.

[20] University of Leicester External Relations Division. How to write a press release. https://www2.le.ac.uk/offices/external/news/publicising/how-to-write-a-press-release.

[21] University of Leicester Press Office. Juices from damaged salad leaves massively stimulate *salmonella* growth and salad leaf colonisation, study shows. University of Leicester Press Office, November 2016. https://www2.le.ac.uk/offices/press/press-releases/2016/november/juices-from-damaged-salad-leaves-massively-stimulate-salmonella-growth-and-salad-leaf-colonisation-study-shows.

[22] C. Wilcox, B. Brookshire, and J. G. Goldman, editors. *Scientific Blogging: The Essential Guide.* Yale University Press, New Haven, 2016.

2.7 Activities

2.7.1 Identifying the Ws in a Press Release

For each of the following excerpts, answer the W-questions (described in Tables 2.1 and 2.2). If it's not possible to answer one, explain why.

- Who are the researchers investigating the problem and conducting the analysis?

- When was the analysis carried out? Answers such as "today", "this week", "recently", etc. are acceptable if the press release does not carry a date.

- Where was the analysis carried out?

- What were the findings from the analysis?

- Why are the findings important?

- Who are the subjects of the study?

- When were the data collected, e.g., when did the subjects participate in the study?

- Where were the data collected, e.g., where were the subjects under study located?

(a) The second paragraph of a press release about a doctor's reassurance appears below.

According to a new study, when a health care provider offers a few encouraging words about their patient's recovery time from an allergic reaction, symptoms are significantly reduced. The researchers, led by Alia Crum, assistant professor of psychology at Stanford University's School of Humanities and Sciences, published the findings recently in the Journal of General Internal Medicine.

(b) A press release that examines the demographics of applicants to charter schools begins with the following two, paragraph-long sentences.

Low-achieving, non-white and poor students stand to gain the most academically from attending charter schools but are less likely to seek charter school enrollment than higher-achieving, more advantaged students who generally live closer to charter schools, according to research from the University of California, Berkeley.

UC Berkeley economist Christopher Walters says in his paper, "The Demand for Effective Charter Schools," forthcoming in the Journal of Political Economy, that examination of charter middle school enrollment in Boston suggests a need to adjust the schools' selection processes by expanding parental outreach and education, and targeting neediest students to better realize charter schools' educational potential while reducing achievement gaps between racial and socio-economic groups.

2.7.2 Reading a Press Release

Visit the media outlet for a research university and find a press release that interests you and that reports on findings from a data-based study. (Alternatively, see Figure 2.3.) Read the press release and complete the following tasks.

- Identify the traditional five Ws in the first paragraph(s) of the press release (see Table 2.1).

- Create an outline of the press release where for each paragraph you write a brief description of the content of the paragraph. Identify whether the paragraph is an introductory paragraph, provides a new detail about the study, supports a detail already given with additional information, supplies background information, or is a concluding paragraph.

- Annotate the outline that you created in the previous step to note whether a paragraph contains a quotation. If it does contain a quotation, indicate whether it is from the investigator(s), another expert, or some other source.

- Use your outline to find three possible places where the press release could be cut to make it shorter. Justify your choice of cut point, and determine whether or not the concluding paragraph(s) can be moved up to remain in the story.

2.7.3 Comparing a News Story and Technical Article

In this activity, you will consider how a news story reports on a technical article. As mentioned in Section 2.2, writing a press release on data findings poses additional challenges specific to the scientific setting.

Find a news story that interests you and that reports on findings from a data analysis. Choose a news story where you can also obtain the technical article on which it is based.

Read both the news story and the technical article. Focus on the methods and data, statistical terms, quotations, and the discussion of limitations, as you answer the following questions.

**Religious upbringing linked to
better health and well-being during early adulthood**

For immediate release: September 13, 2018

Boston, MA – Participating in spiritual practices during childhood and adolescence may be a protective factor for a range of health and well-being outcomes in early adulthood, according to a new study from Harvard T.H. Chan School of Public Health. Researchers found that people who attended weekly religious services or practiced daily prayer or meditation in their youth reported greater life satisfaction and positivity in their 20s—and were less likely to subsequently have depressive symptoms, smoke, use illicit drugs, or have a sexually transmitted infection—than people raised with less regular spiritual habits.

"These findings are important for both our understanding of health and our understanding of parenting practices," said first author Ying Chen, who recently completed her postdoctoral fellowship at Harvard Chan School. "Many children are raised religiously, and our study shows that this can powerfully affect their health behaviors, mental health, and overall happiness and well-being."

The study was published online September 13, 2018 in the *American Journal of Epidemiology*. Previous studies have linked adults' religious involvement to better health and well-being outcomes, including lower risk of premature death.

For this study, Chen and senior author Tyler VanderWeele, John L. Loeb and Frances Lehman Loeb Professor of Epidemiology, analyzed health data from mothers in the Nurses' Health Study II (NHSII) and their children in the Growing Up Today Study (GUTS). The sample included more than 5,000 youth who were followed for between 8-14 years. The researchers controlled for many variables such as maternal health, socioeconomic status, and history of substance abuse or depressive symptoms, to try to isolate the effect of religious upbringing.

The results showed that people who attended religious services at least weekly in childhood and adolescence were approximately 18% more likely to report higher happiness as young adults (ages 23–30) than those who never attended services. They were also 29% more likely to volunteer in their communities and 33% less likely to use illicit drugs.

Those who prayed or meditated at least daily while growing up were 16% more likely to report higher happiness as young adults, 30% less likely to have started having sex at a young age, and 40% less likely to have a sexually transmitted infection compared to those who never prayed or meditated.

"While decisions about religion are not shaped principally by health, for adolescents who already hold religious beliefs, encouraging service attendance and private practices may be meaningful avenues to protect against some of the dangers of adolescence, including depression, substance abuse, and risk taking. In addition, these practices may positively contribute to happiness, volunteering, a greater sense of mission and purpose, and to forgiveness," said VanderWeele.

One limitation of the study is that it consisted mainly of children of white females of relatively high family socioeconomic status, and therefore might not be generalizable to a broader population, though prior research by VanderWeele suggested the effects of religious service attendance for adults may be even larger for black versus white populations. Another limitation was that the study did not look at the influences of parents and peers on adolescents' religious decisions.

While previous studies of adult populations have found religious service attendance to have a greater association with better health and well-being than prayer or meditation, the current study of adolescents found communal and private spiritual practices to be of roughly similar benefit.

Figure 2.3 Press Release on Religious Upbringing.

- Compare the data description and methods section of the technical article to the news story. If the article were one or two paragraphs longer, what additional details would you include in the article? Why do you think it is important to include these details?

- Is there a statistical term that is included in both the news story and the technical article? Are there statistical terms that appear in the article but are left out of the news story? Are there any terms that appear in the article and are translated into non-technical language in the news story? Provide two examples of each, where possible. For each term, explain why you think the journalist chose to include the statistical term, translate it to something non-technical, or omit it entirely from the news story.

- Extract the quotations from the news article. Organize them into three groups: quotations from the author of the article explaining details of the investigation; quotations from an external source adding credibility or a different perspective; human-interest quotations. For each quotation, note how it fits with (or not) some aspect of the technical article.

- Consider the discussion in the technical article about the study limitations. Are these reflected in the news story? Arrange the discussion points in order of importance, and note which ones are emphasized, barely mentioned, or missing entirely in the news story.

2.7.4 Identify Elements of Blog Style

Below are examples that come from the middle of a few blog posts. Highlight the aspects of the writing that follow a blog style. Use the following list to help identify these blog-style elements:

- starting a sentence with "and", "but", or "so" for momentum,

- chatty language rather than formal sentence structure,

- parenthetical interjections,

- mixing content with personal anecdotes or information,

- acknowledgment of potential pitfalls, and

- suggestions for previous or future posts.

(a) As advertised, Flourish is a tool for "powerful, beautiful, easy data visualisation" that lets you "quickly turn your spreadsheets into stunning online charts, maps and interactive stories." Clearly someone on the Flourish team and I are on the same wavelength when it comes to carnival games and data visualization, since the animated rank or relative-score chart that caught my eye turned out to be named the Horserace template.

The hardest part (which wasn't very hard at all) was un-tidying my data to fit their spreadsheet model, with the x-axis variable (number of games played, in my example) as column headers.

The limited "possibility space" makes for a "beginner-friendly" visualization tool. Of course, there's a trade-off–you can't add another layer of labels or annotation with a + sign. Actually, the title in the banner for this post is "after-market" (read: I added it to the image outside of Flourish).

Anyhow, you can see the chart in the embed below, though, if you happen to be on your computer I recommend viewing it here, since my blog template is too narrow a view-port for a chart of this nature (IMHO).

(b) Here the network we trained has given up and guessed the most common color in the image. What makes this question so much harder? Even though the image is cleaner, the question requires many steps of reasoning: rather than simply recognizing the main object in the image, the model must first find the blue cylinder, locate the other object with the same size, and then determine its color. This is a complicated computation, and it's a computation specific to the question that was asked. Different questions require different sequences of steps to solve.

The dominant paradigm in deep learning is a "one size fits all" approach: for whatever problem we're trying to solve, we write down a fixed model architecture that we hope can capture everything about the relationship between the input and output, and learn parameters for that fixed model from labeled training data.

But real-world reasoning doesn't work this way: it involves a variety of different capabilities, combined and synthesized in new ways for every new challenge we encounter in the wild. What we need is a model that can dynamically determine how to reason about the problem in front of it--a network that can choose its own structure on the fly. In this post, we'll talk about a new class of models we call neural module networks (NMNs), which incorporate this more flexible approach to problem-solving while preserving the expressive power that makes deep learning so effective.

(c) For those new to ggplot2, things to watch out for in the code include adding a border round each point (using "stroke") and later on, scaling the points to the range of the Size variable. This is OK because there are only a few points, just don't try it if you've a lot of observations. If you check the plots as they evolve you'll also see the impact of the code in removing the legends and labelling the points directly using the ggrepel package. Until next time.

Part II
Preparing to Write

D ata activities such as cleaning, organizing, exploring, and visualizing are part of the early stages of the data life cycle. This part of the book addresses the preparation of some basic aspects of science writing related to these activities: describing data, creating statistical graphs, and organizing code.

Most well-written science articles contain essential information about the data. A reader needs to know about the data's origin and preparation in order to understand the findings and their implications. The level of detail in the data description depends on the kind of study, the publication venue, and the background of the audience. These issues are the topic of Chapter 3.

An important aspect of communicating with data is the conveyance of an argument through visualization. As with writing, effective graphs require multiple drafts, careful edits that are faithful to the data, and attention to detail. Guidelines for making insightful plots are covered in Chapter 4.

Making the data analysis process open to the scientific community so others can check and confirm our findings is increasingly expected when we publish. Chapter 5 introduces best practices in writing code with the goal of making the analysis computationally reproducible and easy for others to understand.

3   Describing Data

Writing about data is a natural place to start a draft of a technical article. The structure and content of the description are often the same regardless of the angle the article might take so we need not have the story completely figured out before we begin writing, and the goal is well-defined: we want the reader to understand what the data represent, i.e., how the data were collected and prepared for analysis. A well-written description also raises potential limitations of the data that could bear on the conclusions drawn from the analysis.

Typically, a data description provides the source(s) and location(s) where the data can be found, the methods used to collect the data, and a summary of how the data were cleaned and prepared. We address how to write about these aspects in Section 3.1. Then in Section 3.2, we discuss how to describe the variables used in the analysis. Often, a data description also includes simple summary statistics that lay the groundwork for the formal analysis, and in Section 3.3 we provide a few tips on how to write about them. Complete disclosure about the data preparation and exploratory analysis is important for reproducible and replicable science, which is the topic of Section 3.4.

3.1 Data Provenance

The provenance of a dataset traces its origins from raw form to the data actually analyzed. In Chapter 1, we examined how others describe their data and the collection process. Those examples provide helpful templates to follow, but before we begin writing, we also check the kind of data description expected for the venue in which we plan to publish.

We face two basic challenges in describing the data's provenance. The first is that we need to be both thorough and concise. We must sift through the details of the collection and preparation process, select only those most relevant to the analysis, and write concisely about them. A second challenge is to write specific descriptions without

Communicating with Data: The Art of Writing for Data Science. Deborah Nolan and Sara Stoudt, Oxford University Press (2021). © Deborah Nolan and Sara Stoudt. DOI: 10.1093/oso/9780198862741.003.0003

getting into implementation details. For example, the reader does not need to know how information is coded in the source file. All the information they require for a feature is a simple name in plain English and a description of the measurements, such as the units or category descriptors.

3.1.1 Data Collection

When we write about how the data were collected, we consider the following questions:

- When and where did the study take place?

- Who are the subjects under study?

- What procedures were used to select subjects and collect measurements?

- What are the outcomes and covariates?

Additionally, we point out any potential flaws in the data collection. For example, the description might identify possible non-statistical sources of error inherent in the selection and measurement process, or discuss covariates that might be related to the outcome but were not available in the study (particularly if the covariates were available in a related study). Another place to write about these issues is in a discussion section of the article (see Section 7.4.1).

Example: *National Youth Tobacco Survey Description*
The *Morbidity and Mortality Weekly Report* (*MMWR*), published by the US Centers for Disease Control and Prevention (CDC), is a scientific publication that reports on timely public health information and recommendations. The readership of the *MMWR* is primarily health practitioners and researchers. The sixth issue of 2018 contains an article on teen use of e-cigarettes, "Reasons for Electronic Cigarette Use Among Middle and High School Students" by Tsai et al. This 4-page article summarizes findings from the National Youth Tobacco Survey (NYTS), a complex survey carried out by the CDC and the Food and Drug Administration. In the following paragraph from the *MMWR* report, the authors describe the study and its subjects.

NYTS is a school-based, pencil-and-paper questionnaire, self-administered to a cross-sectional, nationally representative sample of students in grades 6-12 in the United States. In 2016, 20,675 students completed the NYTS; the overall survey response rate was 71.6%. Reasons for e-cigarette use were assessed among both ever and current e-cigarette users. Ever users were defined as participants who responded "yes" to the question, "Have you ever used an electronic cigarette or e-cigarette, even once or twice?" Among ever users, current users were those who reported using e-cigarettes on ≥ 1 day during the past 30 days, based on responses to the question, "During the past 30 days, on how many days did you use electronic cigarettes or e-cigarettes?" Current e-cigarette users were further classified into four mutually exclusive groups based on tobacco products used: e-cigarettes only; e-cigarettes and combustible tobacco (e.g., cigarettes, cigars, pipes, bidis, or hookah); e-cigarettes and noncombustible tobacco (e.g., smokeless tobacco, snus, or dissolvable tobacco); and e-cigarettes with combustible and noncombustible tobacco.

The first sentence of this paragraph contains several pieces of information about the survey. For example, we are told that the survey was carried out in middle and high schools using a paper questionnaire. Notice also the precise description of the study group that includes the exact wording of the survey questions along with the response categories. ∎

Example: *A Basketball Player's Record of Hits and Misses*
In the analysis of the hot hand phenomenon (see Section 1.1.1), data were collected on star basketball players. These data are simple strings of 0s and 1s that record each player's shots in a game and whether or not a basket was made. The authors describe the data as follows.

For Curry, Thompson, and Durant, we compiled a string of 1's and 0's representing hits and misses for each of the 99 regular and postseason games that they played in 2016-2017. Curry played 96 games, Thompson played 95, and Durant played 77. We also compiled the strings of hits and misses for the Warrior team, quarter by quarter, leading to 396 = 99 × 4 quarters over the season.

In addition to describing the data collection process, this summary includes information about which players appear in the study and when the games were played. The authors also provide sample sizes, i.e., the number of games recorded for each athlete. ∎

3.1.2 Data Preparation

The data-cleaning phase is a critical part of the analysis pipeline and should be well documented. Data cleaning can reveal potential problems as well as key features of the data, and these revelations can impact the analysis. For these reasons, in technical articles we write about how the data were cleaned and prepared for analysis. The aim is for transparency in describing this process. Topics to address include:

- cases excluded from the analysis,
- quantification and treatment of missing values,
- transformations and derivations of features,
- aggregation of observations to a higher level of granularity, and
- joins of multiple datasets.

Reporting these preparatory actions can help identify potential sources of selection and confirmation bias in the analysis.

Often we summarize this phase of the analysis in a few sentences or paragraphs. Also, we use plots and numerical summaries to highlight unusual or crucial features of the data. We do not write the description as a chronological sequence of our discoveries and data modifications. Instead, we explain and justify the actions taken to prepare the data for analysis.

Example: *Excluded Kenya Donkeys*
In a study of donkeys in Kenya (see Section 1.2.1), three donkeys were excluded from the study.

Three of the 544 donkeys were excluded from the statistical analysis as being unrepresentative: one was a baby, one had a BCS [body condition score] of 1–so was too emaciated–and one, with a BCS of 4.5, was very overweight.

The authors have explained in one sentence the reasons for not including these animals in their analysis. ∎

Example: *Excluded E-Cigarette Users*
In their article on e-cigarettes (see Section 3.1.1), the authors justify dropping one of the four groups of e-cigarette users from their analysis.

Data for the group that used e-cigarettes and noncombustible tobacco products are not presented because of small sample size.

The authors also write about the treatment of missing values.

After excluding participants who had never tried an e-cigarette or had missing information on school level (middle or high), sex, or race/ethnicity (non-Hispanic white, non-Hispanic black, Hispanic, and non-Hispanic other race), 4,049 ever users, including 1,281 current users, were included in the analysis. Data were weighted to account for the complex survey design and adjusted for nonresponse.

This paragraph explains, without going into detail, that student responses are weighted according to the survey design and nonresponse. This statement assures the reader that the analysis was done with careful consideration of the complex design and of missing values in key variables. A technical article might describe the design in greater detail. For example, compare the previous paragraph to the detail provided in the following introductory paragraph from the design section of the survey's associated Methodology Report.

The three-stage cluster sample was stratified by racial/ethnic composition and urban versus rural status at the first (primary) stage. PSUs were classified as "urban" if they are in one of the 54 largest metropolitan statistical areas (MSAs) in the US using 2012 American Community Survey (ACS) data from the US Census

Bureau. Otherwise, they were classified as "nonurban." Additionally, implicit stratification was imposed by geography by sorting the PSU frame by state and by 5-digit ZIP Code (within state). Within each stratum, a PSU was randomly sampled without replacement at the first stage. In subsequent sampling stages, a probabilistic selection of schools and students was made from the sample PSUs.

A more technical article would include a summary similar to this introductory paragraph. However, the *MMWR* article would not include this level of detail. ∎

3.1.3 Multiple Sources

Mashups of multiple data files is an increasingly common technique to create datasets for analysis. Often, new insights can be gleaned from combined data sources that enable more sophisticated analyses, but the descriptions of these joined sources need to clearly explain how the datasets are combined and what an observation corresponds to in the new dataset. That is, we describe, at least conceptually, the level of granularity of the joined data.

Mashups are particularly popular in blog posts. They make terrific examples for teaching others how to extract data from an online resource, use tools from a library or package, and merge data from multiple sources. In the following example, an author laments the unavailability of data at the level of granularity needed so that he could merge other datasets of interest to answer his own questions.

Example: *Blog Opinion Piece on Data Access*
In his post "Data Is A Dish Best Served Raw" on the *eagereyes* blog, Robert Kosara writes about the importance of access to data at a granularity that can be useful for creative analysis. He explains the limitations of sharing data that have been aggregated into tables. Specifically, he mentions a crime report in the UK that answers some interesting questions but doesn't allow further exploration of other questions.

This table, called Who Are The Burglars? appears with others, like What Do They Steal? and How Do They Get In? These are all relevant questions, of course, but what if I want to draw new connections? Perhaps male burglars steal different things than female ones? Do younger people steal more? Is there a difference in how they get in between the different age groups, sexes, and repeat offenders?

None of that is possible with this data, because it's not raw. It has been pre-digested and what we get is a neatly arranged report. There's nothing wrong with a well-prepared report, of course, but providing the source data would make the report a starting point for further exploration, rather than the end product of the analysis.

Kosara praises Data.gov for providing the raw data, which he says has inspired people around the world. He describes several important considerations the UK government is grappling with to make their data available and useful, such as format and location of the data. Kosara takes the position that it's better to have a potentially imperfect solution than to wait for the perfect model, and he concludes:

But despite all the flaws, the inconsistencies, and things that could be done but haven't been, there is one key component that makes it all viable: it's data. Real, raw, original data. Data we can use and melt and recast and analyze. That's the kind of data that is worth going to all the trouble for. If we don't get that kind of data, all the other issues are moot. ∎

3.2 Describing Variables

Readers need information about the variables used in an analysis, and we typically provide only descriptive names and units of measurement. However, there may be situations where there are too many features to describe concisely. In this case, we might organize the variables into groups of similar types of measurements or include only those most relevant to the analysis and list the rest in an appendix or online resource.

3.2.1 Codebooks

Well-documented datasets come with codebooks that provide for each variable: its name and location in the source file; a brief description; summary statistics; missing value codes; censoring codes; and mappings for categories. This detailed information is needed to properly prepare and analyze the data.

However, when we summarize our data for a technical article, we use abbreviated descriptions that are independent of the representation of the information in the data file. For example, rather than include variable names from the codebook, we supply short descriptors, such as "birth weight" not *bwt*, "mother's height" not *mom_ht*, and "smoking status", rather than *smoke*. Relatedly, for categorical data, we include a description of the categories, and not the mapping of categories to their representation as strings or numbers in the source file. The mapping is not relevant to understanding the analysis.

Quantitative variable descriptions should include the units of measurement because they can help the reader interpret the size of an effect or a margin of error. For easy reference, we also suggest providing a table of the variables and their definitions. See Section 3.3 for guidelines for table formatting.

Example: *Infant Health Codebook*
Below is a portion of a codebook for a Child Health and Development Study (CHDS) of infant health. Notice that the codebook in Table 3.1 describes how the categorical features are represented in the file, e.g., never smokers are represented by 0 in the column labeled *smoke*. Also notice that the variable names are abbreviated to make it easy for coding, e.g., *dht* for father's height.

Variable	Definition
bwt	baby's weight at birth in ounces (999 unknown)
parity	total number of previous pregnancies: 0 = first born; 1 = second born; ...; 99 = unknown
ed	mother's education: 0 = less than 8th grade, 1 = 8th to 12th grade – did not graduate; 3 = high school/trade school graduate; 4 = some college; 5 = college graduate; 9 = unknown.
ht	mother's height in inches to the last completed inch: 99 = unknown
dht	father's height, coding same as *ht*
smoke	smoking status of mother: 0 = never; 1 = smokes now; 2 = smoked until current pregnancy; 3 = once did, not now; 9 = unknown

Table 3.1 CHDS Codebook.

In contrast, the following table might appear in a scientific article. Here, the variable names are replaced with one or two word descriptors and the codes for categories and missing data have been eliminated.

Variable	Definition
birth weight	baby's weight at birth in ounces
parity	number of previous pregnancies
mother's education	less than 8th grade; some middle/high school; high school or trade school graduate; some college; college graduate
mother's height	measured in inches to the last completed inch
father's height	measured in inches to the last completed inch
smoking status	smokes while pregnant; smoked until current pregnancy; once smoked, not now; never smoked

Table 3.2 CHDS Codebook Amended.

Table 3.2 contains all the relevant information and reduces the translation work for the reader. ■

3.3 Simple Summaries

When we introduce our data, we usually provide summary statistics, such as a mean or median, and we also provide the associated measure of spread, e.g., a standard deviation (SD) or interquartile range. Sometimes, a visual presentation of summary statistics can be more effective than a table of numbers. Chapter 4 has examples of how to create informative visual data summaries.

Significant Digits We take care with the precision of the statistics we write about. Statistical software usually calculates summary statistics with more digits than needed, and we use the magnitude of the standard deviation and mean to guide us on the number of digits to provide.

Example: *Significant Digits for Average Television Watching*
Consider the following sentence about the mean and standard deviation (SD) of the number of hours of television-watching per day.

The eighth graders watched on average 4.12683 hours of television a day with an SD of 1.33125 hours.

Since the mean is around 4 and the SD is about 1, there is no need to report five digits after the decimal point. Instead, we simply use the values 4.1 and 1.3. No information is lost by dropping the trailing digits because they are irrelevant given the size of the standard deviation. ∎

Formatting Tables When data are displayed in a table, formatting can make a big difference in their readability. Journals and magazines publish guidelines for authors that contain instructions on how to format tables, figures, equations, and headings. We pay attention to these when we begin to write, which saves the hassle and time of reformatting later on.

Example: JASA *Table Formatting Guidelines*
The *Journal of American Statistical Association (JASA)* provides guidelines for authors on how to format tables for publication. These guidelines make it easier for readers to process the numeric information. Below is an excerpt of the guidelines.

In empirical tables it is best to use meaningful abbreviations in row and column labels, rather than let long descriptive labels cause irregular spacing of rows or columns. ...Empirical tables will show better what the numbers are saying if the rows and columns are ordered by some measure of size, rather than being shown in, say, the alphabetical order of the row and column labels. The numbers in the table should be heavily rounded—more than two effective digits are seldom used even by the author in interpretations, so the reader will hardly need them. Row and column averages are also often helpful to the reader.

Notice that these guidelines include a rationale behind the formatting rules. ∎

Report Sample Sizes Sample sizes should also be included in the data description because they give context for interpreting standard error (SE) in a formal analysis. That is, knowing the sample size helps the reader determine whether or not the detected signal is important.

Proportions, Rates, and Ratios These figures measure representation, assess hazards of exposure, and compare proportions (or rates) between groups, respectively. Simple proportions measure prevalence, such as the fraction of a group with a particular characteristic, and when we write a proportion, we need to be clear about both the base (the denominator) and subgroup of the base being counted (the numerator). For example, the description "the proportion of undergraduates who graduated within 5 years of entry to Berkeley" explicitly specifies the base as all students who enter the undergraduate program at a particular university in a specific year. This way, the reader knows exactly who is being counted and who is not.

Example: Proportions of Shots Made
In the article on the hot hand in basketball, Table 3.3 is provided to help the reader understand the structure of the data. We see that each player has one observation per game, i.e., an observation is a string of 0s and 1s that record the success or failure of the attempted shots by the player in a game. The table gives us a sense of the magnitude of the measurements; from the bottom two rows we read that all players had on average 20–24 attempted shots per game with a standard deviation of 5–7 shots.

	Curry	Thompson	Durant	Warriors
Games	96	95	77	99
Observations	96	95	77	396
Season Percentage	56	51	63	56
Average Game Percentage	56	50	62	56
StDev Game Percentage	11	12	12	10
Average Number of Shots	24	20	23	28
StDev Number of Shots	5	5	7	4

Table 3.3 Basketball Data by Player. ∎

Rates Have Units A rate measures change in one quantity per unit change in another quantity, where the second quantity is often time, e.g., the speed of a car in kilometers per hour is an instantaneous rate and the total distance traveled divided by the total travel time is an average rate. Again, the units of both numerator and denominator need to be reported, and when we write about rates, we choose a representation with values that are easy for the reader to interpret, as shown in the following example.

Example: HIV Mortality Rate
In a study of HIV-positive hemophiliacs, the mortality rate is 34.1 deaths per 1,000 person-years; this is an average rate based on the number of deaths and the number of person-years observed in the study period. This study observed 434 deaths among the HIV-positive subjects; the subjects were observed for 12,724 person-years so the

mortality rate is

$$\frac{434 \text{ deaths}}{12,724 \text{ person-years}} = 0.034 \text{ deaths per person-year}$$
$$= 34 \text{ deaths per 1,000 person-years.}$$

The denominator of 1,000 person-years gives a rate that is easier to understand than using a denominator of 1 person-year. ∎

Include Comparisons Typically proportions and rates are not interesting on their own. Instead, we want to know how one compares to another. That way, we can assess whether or not our statistic indicates an interesting difference or a potential problem.

Example: Comparisons of Grade Expectations
A random sample, taken mid-semester, of students in an introductory statistics class found 42% of the 53 male students surveyed expected an A. Alone, this statistic is not that interesting because we do not know if it is an unusually high number or not. We know that the base is the male students surveyed, which should be representative of the class's male students because a random sample was taken. This percentage becomes interesting when compared to the proportion for another group, like the females in the class, or to the proportion of awarded As. In this study, we found that 24% of the 38 females surveyed expected an A and that 20% of the students in the class actually received an A. This additional information makes for interesting comparisons. ∎

Arguments can be based on wrong comparisons, so it is important to be clear about the base. A common mistake occurs in disease tests: the proportion of correctly detected cases among those with the disease is confused with the proportion of true disease cases among those where the test result indicates disease. The denominators of these proportions differ. The first proportion's base is the number with the disease, and the second's is the number of cases that test positive for the disease. Having the disease and testing positive for the disease are different. Positive test results include those with and without disease. When the disease is rare, this confusion can be quite misleading.

Ratios Compare Groups Ratios provide a means of comparing two proportions or two rates. When rates for more than two groups are compared, one group is designated as the reference group. Typically, the reference is the lowest-risk group, and all ratios are calculated with the reference group in the denominator.

Example: Ratio of HIV Mortality Rates
The ratio of mortality rates for HIV-positive and HIV-negative subjects in the study of hemophiliacs (page 59) is

$$\frac{34.1 \text{ HIV-positive d/1,000 py}}{1.6 \text{ HIV-negative d/1,000 py}} = 21.$$

This value indicates that the average mortality rate for HIV-positive hemophiliacs is 21 times that for HIV-negative hemophiliacs. ∎

3.4 Tracking the Analysis

So far in this chapter we have discussed how to write about data, including writing about how we obtained and processed the data. These descriptions help ensure that our results are computationally reproducible, i.e., our work can be computationally recreated from raw data to final findings. (Computational reproducibility is often simply referred to as reproducibility.) We also want our results to be replicable, where another scientist can design and carry out a study and analysis similar to ours and obtain results that confirm our findings. Published work is expected to aim to be both reproducible and replicable, and for this to happen, we need to be clear and thorough in describing the whole analysis pipeline.

3.4.1 Replicable Results

Statistical results are replicable when others can carry out a study similar to ours and achieve consistent results. Ideally, our article has enough information for other researchers to replicate our findings. We must disclose our data preparation and exploration because our decisions in data collection, cleaning, and preparation can impact our findings and these need to be captured in our data description.

A few strategies can help achieve replicability. We can pre-register our analysis. That is, in advance of data collection, we specify the data cleaning, preparation, and formal analysis that we plan to perform. Official websites are available for pre-registration, such as the Open Science Framework. A data description should include a reference to and brief summary of the pre-registration. More generally, we can include a brief discussion of the tests that we plan to do as part of the the data description to ensure readers that we have thought about the best way to use the data to answer the question of interest and that we did not merely happen upon our results after persistent digging.

Another approach is to randomly split the data collected into a test portion and a train portion. The train set is used in the initial exploration and model building; then, the remaining data (the test set) is used to confirm the findings and assess the performance of the fitted model. Information about a test-train split belongs in our data description because it shows how we utilized the data to reach our goals.

Example: *Train and Test Sets of Donkey Data*
In the donkey study (see Section 1.5), the researchers set aside 20% of the observations for use in model assessment. They explain their reasoning for the split and how they divided the data as follows.

For effective treatment it is crucial that we provide a reliable assessment of our tool's accuracy, uncontaminated by our data-driven modelling decisions. Therefore we set aside every fifth case in our dataset after ordering by weight, to be used purely to assess accuracy.

Notice that the authors did not select the test and train sets at random. Instead, they assured the test set has well-distributed inputs by ordering the observations by weight and placing every fifth case in the test set. This process can be reproduced by another researcher interested in checking the results. ∎

3.4.2 Reproducible Wrangling

Our statistical results are computationally reproducible if others can re-compute them from the source data. To achieve reproducibility, we describe the data processing well enough to allow for exact reproduction of the results, beginning with the details of preparing the raw data for analysis through to precise descriptions of the application of statistical methods. Current expectations in the science community are for researchers to provide the code for the entire data-analysis process. Reproducible coding practices are described in Chapter 5. In addition to providing code, the written data description should provide information about the main steps in data handling.

3.4.3 Accessible Data

In order for others to reproduce our analysis and have a clear understanding of how to replicate our work, we sometimes need to do more than describe the data. For example, we might grant access to our data. Scientific societies and organizations have developed policies for sharing data that authors need to abide by in order to publish in their journals. For example, the editorial policy for *Science* insists that "All data used in the analysis must be available to any researcher for purposes of reproducing or extending the analysis." The *Science* policy describes several ways to make data available, including:

- published in the paper, if the data are small enough;

- accessible as supplementary material in a journal repository, for medium-sized data;

- deposited in a special-purpose repository, such as the Image Data Resource, or a general-purpose database, for large data; and

- archived at a general repository as a static version, for dynamic data.

Below, we provide examples of how authors have made their data available.

Example: *Small Data: Planck Constant*
The article about the accuracy, consistency, and stability of measurements of the Planck constant performs a meta-analysis (see Section 1.2.2). In a meta-analysis, the results from other studies are collected and analyzed; that is, the data are statistics from other studies. This paper on the Planck constant includes the metadata as a table of summary statistics from the various studies. ∎

Example: *Medium Data: Donkeys*
The authors of the article on donkeys in Kenya (see Section 1.2.1) have made their data available in an *R* package, *paranomo*. A link to this *R* package appears on Jonathan Rougier's web page. It is common for scientists who have developed novel methods for analyzing data to prepare software for distribution and include their data in the software. ∎

Example: *Large Data: National Youth Tobacco Survey*
The e-cigarette article (see Section 3.1.1) makes use of data from a large federal survey, the 2016 National Youth Tobacco Survey (NYTS). The survey data are available on a website maintained by the CDC. The data description in the article includes a reference to the NYTS site where the data can be found. ∎

Example: *Large Data: Light Curves*
Astronomer Helen A. C. Giles discovered an exoplanet by analyzing data from the K2 satellite telescope. In the write-up of her findings (see Section 1.3.1), she briefly describes the source of the data:

EPIC248847494b was observed in Campaign 14 of the K2 mission in long cadence (29.4 minute exposures). The campaign began on 1st June 2017 at 05:06:29 UTC and ended 19th August 2017 at 22:11:02 UTC, lasting 79.7 days.

These two sentences contain enough information to find this publicly available data. On the K2 website, we find additional information about how to access the data. The K2 mission is an example of the terrific possibilities that can be obtained by making data and documentation publicly available. ∎

Unfortunately at times, privacy issues do not allow data products to be openly available. In this case, the scientific community has developed protocols to vet researchers and share resources beyond the original circle of scientists.

3.5 Notes

Many of the example data descriptions in this chapter come from articles presented in Chapter 1. These include the articles on hot hands by Daks, Desai, and Goldberg [4], Kenyan donkeys by Milner and Rougier [12], the redefinition of the kilogram by Possolo et al. [18], and exoplanets by Giles et al. [7]. The data for the Milner and Rougier study are available in their *R* package [20], and the exoplanet data can be found at the Space Telescope Science Institute site [22]. In addition, the e-cigarette example comes from an article in the Morbidity and Mortality Weekly Report by Tsai et al. [23] and the more technical methodology report on the National Youth Tobacco Survey [16]. These data are available from the Centers for Disease Control and Prevention [2]. Excerpts from Robert Kosare's blog [10] address questions around data sharing.

The data for the CHDS and HIV mortality studies, the exit poll, and grade-expectation survey (and their codebooks) are described in greater detail in Nolan and Speed [14] and are available online at the StatLabs site [13]. Originally, the HIV mortality data came from the *Multicenter Hemophilia Cohort Study*, sponsored by the US National Cancer Institute and described in Goedert et al. [8]. The Child Health and Development Studies was a comprehensive investigation of all pregnancies that occurred between 1960 and 1967 among women who received prenatal care in the Kaiser Foundation Health Plan in the San Francisco–East Bay area and delivered at one of the Kaiser hospitals in northern California (see Yerushalmy [24]). The exit poll was conducted by the Field Research Corporation and is described in Freedman et al. [5]. The data descriptions in Section 3.7.1 are based on student reports about the exit poll.

The example on significant digits and Sections 3.7.2, 3.7.3, and 3.7.4 use data from the Early Childhood Longitudinal Study [15]. Additionally, the statistics in Activity 3.7.4 are based on those found in Cunningham [3].

For general information on replicable and reproducible research, see Gelman & Loken [6], Leek & Peng [11], and Peng [17]. Information about pre-registration on the Open Science Framework is available at [1]. The rOpenSci Reproducibility in Science guide [19] has many additional references.

The editorial policy for *Science* can be found at [21]. The *Journal of the American Statistical Association*'s style guide can be found at [9].

For more practice, consider working on Portfolio Piece A.

3.6 References

[1] Center for Open Science. Open science framework. https://osf.io/, 2020.

[2] Centers for Disease Control and Prevention. National Youth Tobacco Survey (NYTS). https://www.cdc.gov/tobacco/data_statistics/surveys/nyts/.

[3] S. A. Cunningham, M. R. Kramer, and K. M. V. Narayan. Incidence of childhood obesity in the United States. *New England Journal of Medicine*, 370:403–411, 2014.

[4] A. Daks, N. Desai, and L. R. Goldberg. Do the Golden State Warriors have hot hands? *The Mathematical Intelligencer*, 40(4):1–5, 2018.

[5] D. A. Freedman, S. P. Klein, J. Sacks, C. A. Smyth, and C. G. Everett. Ecological regression and voting rights. *Evaluation Review*, 15(6), 1991.

[6] A. Gelman and E. Loken. The statistical crisis in science. *American Scientist*, 102(6):460–466, 2014.

[7] H. A. C. Giles, H. P. Osborn, S. Blanco-Cuaresma, C. Lovis, D. Bayliss, P. Eggenberger, A. C. Cameron, M. H. Kristiansen, O. Turner, F. Bouchy, and S. Udry. Transiting planet candidate from K2 with the longest period. *Astronomy & Astrophysics*, 615:L13,1–5, 2018. arXiv:1806.08757v1.

[8] J. J. Goedert, C. M. Kessler, L. M. Aledort, R. J. Biggar, W. A. Andes, G. C. White, J. E. Drummond, K. Vaidya, D. L. Mann, M. E. Eyster, M. V. Ragni, M. M. Lederman, A. R. Cohen, G. L. Bray, P. S. Rosenberg, R. M. Friedman, M. W. Hilgartner, W. A. Blattner, B. Kroner, and M. H. Gail. A prospective study of human immunodeficiency virus Type 1 infection and the development of AIDS in subjects with hemophilia. *New England Journal*

of Medicine, 321:1141–1148, 1989.

[9] *Journal of the American Statistical Association*. American Statistical Association style guide. http://amstat.tfjournals.com/asa-style-guide/.

[10] R. Kosara. Data is a dish best served raw. https://eagereyes.org/data/dish-best-served-raw, June 2009.

[11] J. T. Leek and R. D. Peng. Opinion: Reproducible research can still be wrong: Adopting a prevention approach. *Proceedings of the National Academy of Sciences of the United States of America*, 112(6):1645–1646, 2015.

[12] K. Milner and J. Rougier. How to weigh a donkey in the Kenyan countryside. *Significance*, 11(4):40–43, 2014.

[13] D. Nolan and T. Speed. Maternal smoking and infant health II. https://www.stat.berkeley.edu/users/statlabs/labs.html#babies, 2000.

[14] D. Nolan and T. Speed. *Stat Labs: Mathematical Statistics Through Applications*. Springer, New York, 2000. https://www.stat.berkeley.edu/users/statlabs/labs.html.

[15] Office on Smoking and Health. Early Childhood Longitudinal Study, Kindergarten Class of 1998–99 (ECLS-K), Combined User's Manual for the ECLS-K Eighth-Grade and K–8 Full Sample Data Files and Electronic Codebooks (NCES 2009–004). National Center for Education Statistics, Institute of Education Sciences, 2009.

[16] Office on Smoking and Health. 2018 National Youth Tobacco Survey: Methodology report. Technical report, U.S. Department of Health and Human Services, Centers for Disease Control and Prevention, National Center for Chronic Disease Prevention and Health Promotion, Office on Smoking and Health, 2018.

[17] R. Peng. Reproducible research in computational science. *Science*, 6060:1226–1227, 2011.

[18] A. Possolo, S. Schlamminger, S. Stoudt, J. R. Pratt, and C. J. Williams. Evaluation of the accuracy, consistency, and stability of measurements of the Planck constant used in the redefinition of the international system of units. *Metrologia*, 55:29–37, 2018.

[19] rOpenSci. Reproducibility in science: A Guide to enhancing reproducibility in scientific results and writing. http://ropensci.github.io/reproducibility-guide/sections/introduction/.

[20] J. Rougier. Jonathan (Jonty) Rougier's homepage. https://people.maths.bris.ac.uk/~mazjcr/.

[21] *Science*. *Science* journals: editorial policies. http://www.sciencemag.org/authors/science-journals-editorial-policies, 2019.

[22] Space Telescope Science Institute. K2sff. https://archive.stsci.edu/prepds/k2sff/.

[23] J. Tsai, K. Walton, B. N. Coleman, S. R. Sharapova, S. E. Johnson, S. M. Kennedy, and R. S. Caraballo. Reasons for electronic cigarette use among middle and high school students-National Youth Tobacco Survey, United States, 2016. *Morbidity and Mortality Weekly Report*, 67(6):196–200, 2018.

[24] J. Yerushalmy. Mother's cigarette smoking and survival of infant. *Am. J. Obstet. Gynecol.*, 88:505–518, 1964.

3.7 Activities

3.7.1 Editing Data Descriptions

Below are two attempts to describe the same dataset used to study voting behavior. Read each description and make a list of its pros and cons. Then, write your own data description that improves upon the pros of these examples and addresses the cons.

Version A: I conducted an analysis of the results from an exit poll for the 1988 Democratic presidential primary. The sampling method for voter selection was a simple random sample. Below is a description of the data:

Variable	Description
Precinct	Identification number for the precinct (neighborhood)
Candidate	Vote cast for candidate: 1 = Jackson; 2 = LaRouche; 3 = Dukakis; 4 = Gore; 5 = Simon
	Jackson is the only black candidate (the other four were White).
Race	Voter's race: 0 = missing; 1 = White; 2 = Hispanic; 3 = Black; 4 = Asian; 5 = other
Income	Voter's annual income (thousands of $) 0 = missing; 1 = 00–10; 2 = 10–20; 3 = 20–30; 4 = 30–40; 5 = 40–50; 6 = 50–60; 7 = 60–70; 8 = 70+

Table 3.4 Voting Data.

The dataset in Table 3.4 is in matrix format, with each row representing a single voter. Rows containing missing information in race and income are removed (8.357516% of all rows). Precinct, race and income are explanatory variables with categorical levels, and candidate is the response variable with categorical levels.

Version B: The dataset analyzed comes from an exit poll taken by the Field Research Corporation, a private research firm performing consulting work for government interests. The poll surveyed 1,867 voters on race, income level, precinct of residence, and candidate supported. Five racial groups, coded 1–5, are White, Hispanic, Black, Asian, and other. Eight income groups are coded 1-8 and represent $0–10k, $10–20k, $20–30k, $30–40k, $40–50k, $50–60k, $60–70k, and $70k+ annual income. Thirty-nine precincts are represented. The data form a simple random sample of a small city population.

3.7.2 Describing a Complex Survey

Below is an excerpt from the manual for the Early Childhood Longitudinal Study, Kindergarten Class of 1998–1999 (ECLS-K). Write a one- to two- paragraph summary of ECLS-K that would be appropriate for an *MMWR* on childhood obesity (see Section 3.1.1 for a guide).

This manual provides guidance and documentation for users of the eighth-grade data of the Early Childhood Longitudinal Study, Kindergarten Class of 1998–1999 (ECLS-K). It begins with an overview of the ECLS-K study. Subsequent chapters provide details on the instruments and measures used, the sample design, weighting procedures, response rates, data collection and processing procedures, and the structure of the data file.

The ECLS-K focuses on children's early school experiences beginning with kindergarten and ending with eighth grade. It is a multi-source, multi-method study that includes interviews with parents, the collection of data from principals and teachers, and student records abstracts, as well as direct child assessments. In the eighth-grade data collection, a student paper-and-pencil questionnaire was added. The ECLS-K was developed under the sponsorship of the U.S. Department of Education, Institute of Education Sciences, National Center for Education Statistics (NCES). Westat conducted this study with assistance provided by Educational Testing Service (ETS) in Princeton, New Jersey.

The ECLS-K followed a nationally representative cohort of children from kindergarten into middle school. The base-year data were collected in the fall and spring of the 1998–99 school year when the sampled children were in kindergarten. A total of 21,260 kindergartners throughout the nation participated.

Two more waves of data were collected in the fall and spring of the 1999–2000 school year when most, but not all, of the base-year children were in first grade. The fall-first grade data collection was limited to a 30 percent subsample of schools (see exhibit 1-1). It was a design enhancement to enable researchers to measure the extent of summer learning loss and the factors that contribute to such loss and to better disentangle school and home effects on children's learning. The spring-first grade data collection, which included the full sample, was part of the original study design and can be used to measure annual school progress and to describe the first-grade learning environment of children in the study. All children assessed during the base year were eligible to be assessed in the spring-first grade data collection regardless of whether they repeated kindergarten, were promoted to first grade, or were promoted to second grade. In addition, children who were not in kindergarten in the United States during the 1998–99 school year, and therefore did not have a chance to be selected to participate in the base year of the ECLS-K, were added to the spring-first grade sample. Such children include immigrants to the United States who arrived after fall 1998 sampling, children living abroad during the 1998–99 school year, children who were in first grade in 1998–99 and repeated it in 1999–2000, and children who did not attend kindergarten. Their addition allows researchers to make estimates for all first-graders in the United States rather than just for those who attended kindergarten in the United States in the previous year.

A fifth wave of data was collected in the spring of the 2001–02 school year when most, but not all, of the sampled children were in third grade. In addition to the school, teacher, parent, and child assessment data collection components, children were asked to complete a short self-description questionnaire, which asked them how they thought and felt about themselves both academically and socially. The spring-third grade data collection can be used to measure school progress and to describe the third-grade learning environment of children in the study. ...

The sample of children in the eighth-grade round of data collection of the ECLS-K represents the cohort of children who were in kindergarten in 1998–99 or in first grade in 1999–2000. Since the sample was not freshened after the first-grade year with children who did not have a chance to be sampled in kindergarten or first grade (as was done in first grade), estimates from the ECLS-K eighth-grade data are representative of the population cohort rather than all eighth-graders in 2006–07. Comparisons of the weighted population of ECLS-K children enrolled in the eighth grade with the weighted population of eighth-graders reported in the 2006 Current Population Survey suggest that the ECLS-K represents about 80 percent of all U.S. eighth-graders in the 2006–07 school year.10 Some examples of subpopulations of eighth-graders who are not represented in the ECLS-K in 2006–07 include children who started kindergarten before fall of 1998 and were retained in a later grade, children who immigrated to the United States after first grade, and children who were home-schooled until after first grade. Data were collected from teachers and schools to provide important contextual information about the school environment for the sampled children, but the teachers and schools are not representative of eighth-grade teachers and schools in the country in 2006–07. For this reason, the only weights produced from the study for eighth-grade estimates are for making statements about children, including statements about the teachers and schools of those children.

The ECLS-K has several major objectives and numerous potential applications. The ECLS-K combines (1) a study of achievement in the elementary and middle school years; (2) an assessment of the developmental status of children in the United States at the start of their formal schooling and at key points during elementary and middle school; (3) cross-sectional studies of the nature and quality of kindergarten programs in the United States; and (4) a study of the relationship of family, preschool, and school experiences to children's developmental status at school entry and their progress during kindergarten, elementary school, and middle school.

The ECLS-K has both descriptive and analytic purposes. It provides descriptive data on children's status at school entry, their transition into school, and their progress into middle school. The ECLS-K also provides a rich dataset that enables researchers to analyze how a wide range of family, school, community, and individual variables affect children's early success in school; to explore school readiness and the relationship between the kindergarten experience and middle school performance; and to record children's academic growth as they move into middle school.

The ECLS-K is part of a longitudinal studies program comprising two cohorts: kindergarten cohort and a birth cohort. The birth cohort (ECLS-B) is following a national sample of children born in the year 2001 from birth to kindergarten. The ECLS-B examines how early learning environments are associated with early cognitive, physical, and socioemotional development and thus prepare children for kindergarten success. Together these cohorts will provide the depth and breadth of data required to more fully describe and understand children's early learning, development, and education experiences.

3.7.3 Translating a Codebook into a Data Description

Table 3.5 below contains a small portion of the codebook for the Early Childhood Longitudinal Study, Kindergarten Class of 1998–99 (ECLS-K). Write a one-paragraph description of these variables that would be appropriate for an MMWR report on obesity. Also, create a table of these variables that might appear in the MMWR report. Use Section 3.2.1 for a guide.

ID	Name	Category	Derived from	File values
1	*R7AGE*	Child	*R7DOBMM* *R7DOBDD* *R7DOBYY*	1 = less than 126; 2 = 126 to less than132; 3 = 132 to less than 138; 4 = 138 to less than 144; 5 = 144 or more
2	*R7GENDER*	Child	*R6GENDER* *CHILDGEN* *GENDER*	1 = Male; 2 = Female
15	*C7BMI*	Child	*C7HEIGHT* *C7WEIGHT*	Continuous
16	*C7HEIGHT*	Child	*C7HGT1* *C7HGT1A* *C7HGT2*	Continuous in inches to nearest quarter inch
17	*C7WEIGHT*	Child	*C7WGT1* *C7WGT1A* *C7WGT2*	Continuous in pounds to nearest half inch
40	*W8SESQ5*	Family/HH	*W8SESL*	1 = First quintile (lowest); 2 = Second quintile; ... ; 5 = Fifth quintile (highest)
60	*P7FSSTAT*	Family/HH	*P7WORRFD* *P7FDLAST* *P7BLMEAL* *P7CUTML* *P7HUNGRY*	1 = Food secure; 2 = Food insecure without hunger; 3 = Food insecure with hunger (moderate); 4 = Food insecure with hunger (severe)

Table 3.5 ECLS-K Codebook.

3.7.4 Writing about Summary Statistics

Below are summary statistics from the Early Childhood Longitudinal Study that pertain to obesity. These are reported as the return values from statistical software calculations.

Prevalence is the proportion of all children in an age group who are obese. Below are figures for the prevalence of obesity in children in kindergarten, first, third, fifth, and eighth grades.

0.1239224 0.1261889 0.1858205 0.2190232   0.2083214

The standard errors for the above prevalence estimates of obesity in children in kindergarten, first, third, fifth, and eighth grades appear below.

0.006339274 0.011746849 0.006337093 0.007689132 0.008314215

The incidence rate (cases per person-year) of obesity in eighth-grade children who were normal weight in kindergarten, according to socio-economic status (first, second, ..., fifth quintile) appear below.

0.020617843 0.023140108   0.022495601   0.012335266 0.009986594

Also, included here are the incidence rates of obesity in eighth-grade children who were overweight in kindergarten, according to the quintiles of socio-economic status.

0.07995644 0.08665081 0.11441536 0.10973347   0.07167146

Finally, provided below are the risk ratios of incidence for overweight versus normal-weight children, according to the quintile of socio-economic status.

3.384671 3.175551 3.857810 5.070478 4.990220

For each of the following prompts, write a sentence that uses some of the figures from above. Make sure that you round the numeric values to a reasonable level.

- Compare the prevalence of obesity in kindergarten and eighth grade. Use percentages in this comparison.

- Rewrite the above sentence to include standard errors along with percentages.

- Describe the incidence rate of obesity in a child who was normal weight in kindergarten. The rate should be expressed in cases per 1,000 person-years.

- Calculate the risk ratio of the incidence of obesity for overweight and normal-weight children in one socio-economic status group. Write a sentence that explains this risk ratio.

- Compare the risk ratio for two socio-economic status groups and relate the difference between these groups to the incidence of obesity.

- Organize into a table the incidence rates of obesity for normal and overweight children and the risk ratios for these two groups (normal and overweight), according to socio-economic status. Consider which variable should correspond to rows and which to columns in your table. Take care with the number of digits provided in the table.

4 Communicating Through Statistical Graphs

Technical reports include plots which must be "read" along with the text. A statistical graph should add convincing support to a written argument or offer an easy-to-digest presentation of key numerical summaries. In this chapter, we consider how to create plots that are effective in communicating statistical findings. We address how to select an appropriate type of plot to reveal underlying structure in the data, facilitate important comparisons, and create a context for interpreting the distributions and relationships observed. Our discussion focuses on topics similar to those introduced in Chapter 1. We consider how to read standard univariate and bivariate plots, pay attention to the details in making and interpreting plots, and create plots that help convince the reader of our argument.

4.1 Matching Plot to Data

The type of data we want to plot can be a good indicator for the kind of plot to make. Table 4.1 gives a mapping of the various plots that are typically good options for each type of data. Whether the variable(s) are quantitative or qualitative generally determines the set of viable plots, although there are exceptions. Other factors that enter into the decision are the number of observations, and whether the data takes on only a few distinct values (for example, the distinction between bar plots and histograms).

4.2 Reading Plots

A first step in learning how to create a good plot is to be able to read plots. We provide examples of reading plots from one study: the Child Health and Development Study (CHDS) (see Section 3.5). These data are for 1,236 babies–baby boys born during one

Communicating with Data: The Art of Writing for Data Science. Deborah Nolan and Sara Stoudt, Oxford University Press (2021). © Deborah Nolan and Sara Stoudt. DOI: 10.1093/oso/9780198862741.003.0004

Data Type	Plot
	Univariate
Quantitative	Rug plot, histogram, density curve, box-and-whisker plot, violin plot, normal quantile plot
Qualitative	Bar plot, dot chart, line plot, pie chart
	Bivariate
Both Quantitative	Scatter plot, smooth curve, contour plot, heat map
Both Qualitative	Side-by-side bar plots, mosaic plot, overlaid lines
Quantitative/Qualitative	Overlaid density curves, side-by-side box-and-whisker plots, overlaid smooth curves, quantile-quantile plot

Table 4.1 Plot Types for Different Data Types. The choice of plot to display a univariate distribution or bivariate relationship depends in part on the type of data. Additional considerations, such as the number of observations, help determine the most appropriate choice.

year of the study who lived at least 28 days and were single births (e.g., not one of a twin). The variables available for making plots are provided in Table 4.2; the variety of data types is useful for demonstrating how to read many different kinds of plots. Statistical graphs are valuable for presenting all kinds of data, not just observational studies like the CHDS; that is, visual displays of metadata, summary statistics, simulation-study results, and model output can make effective presentations. See for example Activities 4.8.3 and 4.8.3 at the end of the chapter.

Variable	Definition
Gestation	Length of pregnancy in days
Birthweight	Baby's weight at birth in ounces
Parity	Number of previous pregnancies including fetal deaths and stillbirths
Smoking status	Smokes now (while pregnant); smoked until current pregnancy; once smoked, not now; never smoked
Education	Highest level of education: less than 8th grade, some middle/high school 12th grade; high school graduate; trade school; some college; college graduate (or higher).
Mother's height	Measured in inches to the last completed inch
Mother's weight	Pre-pregnancy weight in pounds
Father's height	Measured in inches to the last completed inch
Father's weight	Measured in pounds

Table 4.2 Data Dictionary for the Child Health and Development Study.

4.2.1 Reading Distributions

When interpreting a histogram or density curve, we examine: the symmetry and skewness of the distribution; the number, location, and size of high-frequency regions (modes); the length of tails (often in comparison to the normal curve); gaps where no values are observed; and unusually large or anomalous values. Figure 4.1 provides an example of a distribution with several of these features. When we read a distribution, we connect the features that we see in the plot to the quantity measured. For example, the distribution of the parity of the pregnancy in the CHDS study, appears in Figure 4.2. A value of zero means this baby is the mother's first pregnancy, one corresponds to her second pregnancy, etc. From the histogram, we see that the distribution of parity is unimodal with a peak at zero. Nearly the same proportions of women are in their first or second pregnancy. We also see that the distribution is heavily skewed right, with a long right tail indicating that some few mothers have between 6 and 13 previous pregnancies. Although quantitative, mother's parity is discrete because only integer values are possible. For this reason, the bins are centered on the integers, e.g., the bin from 1.5 to 2.5 contains only those mothers with parity two. The rightmost bins of the histogram are wider, e.g., the mothers with parity six and seven are combined into one bin, and we compute the corresponding percentage as 2 (width) × 0.02 (height) or about 4%. (When bin counts are small, we use wider bins to smooth the distribution because we do not want to read too much into the fluctuations of small numbers.)

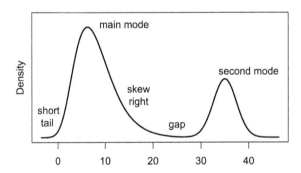

Figure 4.1 Example Density Curve. The distribution shown here has two modes, with the main mode around 8 and a smaller mode near 35. The main mode is skewed right. The gap between the two modes indicates two distinct groups of subjects.

Bar Plot ≠ Histogram With qualitative data, the bar plot serves a similar role as the histogram. That is, the bar plot gives a visual presentation of the "popularity" or frequency of different groups. However, we cannot interpret the shape of the bar plot in the same way as a histogram, e.g., tails and symmetry do not make sense in this setting. Also, the frequency of a category is represented by the height of the bar; the

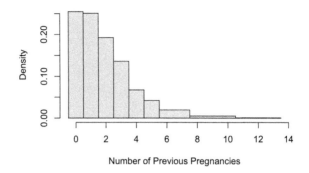

Figure 4.2 Histogram of Parity. Parity, the number of previous pregnancies, has a discrete distribution because only integer values are possible. The histogram reveals the distribution is unimodal, skewed right, with a long right tail where a few women in the study had between 6 and 13 previous pregnancies.

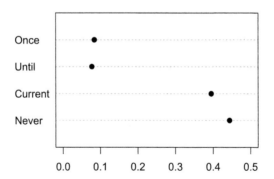

Figure 4.3 Distribution of Mother's Smoking Status. This dot chart displays the proportion of mothers in each smoking status. The labels denote whether the mother "Never" smoked, "Current"ly smokes (while pregnant), smoked "Until" she became pregnant, or smoked "Once" and quit before pregnancy. The dot chart uses length to represent the proportion of each type of smoker. We see that many mothers (about 40%) smoked during their pregnancy and roughly the same number (about 44%) have never smoked.

width carries no information about the distribution. In the extreme, the dot chart in Figure 4.3 eliminates the bars entirely and represents each percentage by a single dot. Reading this plot, we see there are roughly the same proportion of CHDS mothers who smoked during pregnancy (40%) as those who never smoked (44%).

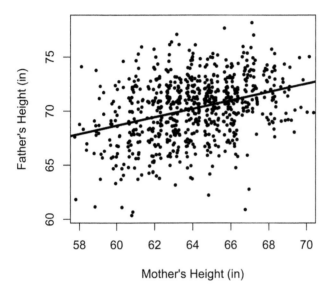

Figure 4.4 Heights of Mothers and Fathers. This scatter plot shows a weak linear association between the heights of mothers and fathers in the CHDS. According to the fitted line, a mother who is 64 in tall (the average height for the mothers) tends to have a partner 70.2 in tall, on average, and a mother 2 in taller (66 in) tends to have a partner 71 in tall, on average. The correlation is 0.34, indicating the variability in father's height for a group of mothers who have the same height, is quite large.

4.2.2 Reading Relationships and Trends

When we investigate multiple variables, we examine the relationship between them in addition to their univariate distributions. If both variables are quantitative, then we usually examine their relationship with a scatter plot. We look for linear and simple nonlinear relationships, i.e., where a transformation of one or the other or both variables leads to a linear relationship.

Figure 4.4 displays a scatter plot of mother's and father's height (both are quantitative variables). We observe that mothers who are above average in height tend to have partners who are also above average in height, and the relationship appears linear.

However, the association between mother's and father's height is weak as there are many exceptions to this "rule", indicated by the large spread of the points about the line on the plot.

Two Univariate Plots ≠ One Bivariate Plot The histograms for two quantitative variables do not contain enough information to create their scatter plot so we must exercise caution when we read a pair of histograms. That is, the histograms do not show how these two quantities vary together. We need to examine a scatter or line plot to get a sense of the relationship between two quantitative variables.

Transformations When we have a simple nonlinear relationship, we look for transformations of the x and/or y variables that straighten the relationship because it is easier for us to discern a linear relationship than to differentiate between curvilinear relationships. Figure 4.5 gives a simple example of this. The left scatter plot reveals a curvilinear relationship between x and y, but we do not know whether it is, for example, log-linear or log-log. The middle plot reveals a different curvilinear relationship between $\log(y)$ and x; this plot indicates that the relationship is not log-linear relationship. A further transformation (in the rightmost plot) of $\log(y)$ against $\log(x)$ establishes a log-log relationship because the scatter of the transformed points looks linear.

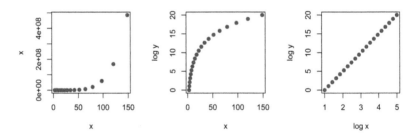

Figure 4.5 Straightening Relationships. These scatter plots demonstrate how transformations that straighten a bivariate relationship can be used to uncover the original relationship between two variables. The plots are (x,y) on the left, (x, log(y)) in the middle, and (log(x), log(y)) on the right. Since the points on the right fall on a line, x and y have a log-log relationship.

One Qualitative and One Quantitative Variable When we examine the relationship between a quantitative and a qualitative variable, we often use the qualitative variable to divide the data into groups and compare the distribution of the quantitative variable across these groups. For example, we can compare the distribution of birthweight for babies born to mothers with different smoking statuses (Figure 4.6). We see that the distribution of birthweight for the current and never smokers are similar in shape, but the distribution of birthweight for babies born to never smokers is shifted to the right by about eight ounces. We also see that there is less variability in birthweight for never smokers.

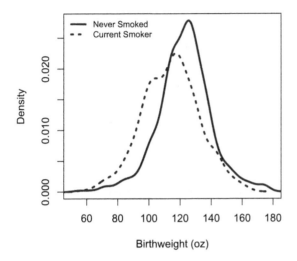

Figure 4.6 Distribution of Birthweight by Mother's Smoking Status. This plot compares the distribution of birthweight for two groups of mothers. The density curve of birthweight for babies born to mothers who never smoked (solid curve) is overlaid on the plot of the density curve for mothers who smoked during pregnancy (dashed curve). The two distributions are similar in shape, except mothers who never smoked have babies that weigh about eight ounces more on average than babies born to current smokers. The birthweight of never smokers also appears more tightly distributed about its center.

Two Qualitative Variables With two qualitative variables, we often examine the distribution of one variable across subgroups defined by the second variable. In effect, we condition on one variable and plot the distribution of the second. For example, we might consider the distribution of never and current smokers given different levels of education. In general, more highly educated mothers tend to live healthier lives which can impact the health of the newborn, and we want to look for the presence of this relationship in these data. We can compute the proportion of mothers in each smoking status within an education level and display it in a table.

In Table 4.3, we have conditioned on education level so each column sums to one. Figure 4.7 displays a line plot of these proportions. There is one line (set of connected dots) for each education level. We see that less-educated women have higher rates of smoking while pregnant.

		Education			
		<HS	HS	Some Col	College
	Never	0.35	0.46	0.47	0.50
Smoking Status	Current	0.53	0.40	0.35	0.30
	Until	0.07	0.09	0.07	0.08
	Once	0.06	0.05	0.12	0.12

Table 4.3 Smoking and Education.

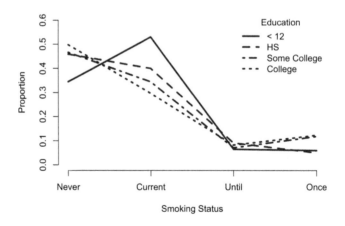

Figure 4.7 Line Plot of Mother's Smoking Status and Education. This line plot compares smoking status across education levels. Each connected line displays the proportion of mothers with a particular smoking status within an education level, e.g., the proportions of high-school-educated mothers (connected by dashed lines) add to one across the four smoking statuses. Most notable is that the college-educated mothers are much less likely to smoke during their pregnancy (30%) than those with a high school education (40%) or less (53%).

4.3 Comparisons in Multivariate Settings

When we examine a distribution or relationship, we often want to compare it across subgroups of the data. This process of conditioning on additional factors often leads to visualizations that involve three or more variables.

Qualitative variables naturally divide the data into subgroups, and we often want to examine the distribution of quantitative variables across these subgroups. Figure 4.8 is an example where we compare the relationship between baby's birthweight and mother's height for never and current smokers. The points in the scatter plot have been coded according to whether the mother never smoked (diamond) or currently smokes (circle). We have added two curves to this plot: a dashed line for smokers and solid line for never smokers. As with the scatter plot of mother's and father's height, we find a weak linear association. For each group of mothers (never or current smokers) the curve displays the average birthweight of the babies born to mothers with roughly the same height. One interesting feature we see is that both curves are roughly linear and parallel, with the curve for smokers about eight ounces lower than for never smokers. In other words, for mothers of a particular height, the babies born to smokers tend to be weigh approximately eight ounces less on average than those born to never smokers.

Below are some plotting techniques to help make comparisons and uncover structure. They include two- and three-variable plots of quantitative and categorical data.

Combinations Across Groups When we examine relationships between qualitative variables, we examine proportions of one variable within subgroups defined by other variable(s). Figure 4.7 shows such a comparison between two qualitative variables. With three (or more) qualitative variables, we can subdivide the data according to the combinations of levels of these variables and compare proportions with line plots, dot charts, side-by-side bar charts, mosaic plots, etc. We typically organize the lines, dots, and bars into groups according to a combination of levels from two (or more) variables.

Example: Weekly Earnings for Men and Women in 2014
The Current Population Survey (CPS) tracks income statistics for the working population of the USA for full-time workers who are over 25. Figure 4.9 displays the CPS estimates for median weekly income by sex and education. The two line plots help us compare the gap in income between men and women and see how this gap grows with education level. ∎

Panels of Scatter Plots With quantitative data, we often create scatter plots of all pairs of variables. However, these plots of pairs may not be able to capture the high-dimensional relationship between the variables. Studying relationships between three or more variables can be difficult. One technique for examining these multivariable relationships is to make a grid of conditional plots. For example, we often want to know whether a relationship between x and y is the same for different values of z. We can divide the data into subgroups according to the value of z; that is, in each subgroup, the value of z is roughly constant. We make a grid or ensemble of scatter plots to compare

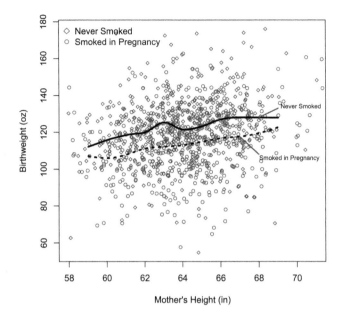

Figure 4.8 Scatter Plot of Baby's Birthweight and Mother's Height. Mother's height and baby's birthweight have a weak linear relationship with a correlation coefficient of about 0.20. The points in this scatter plot are coded according to the mother's smoking status (diamond for never smoked and circle for smoked during pregnancy). Two curves added to this plot show the average birthweight for babies born to mothers with the same height (dashed for mothers who smoked during pregnancy and solid for mothers who never smoked). The relationship between birthweight and mother's height is roughly linear for both groups of mothers and parallel, where babies born to never smokers consistently weigh 8--10 oz more on average than those born to current smokers.

the relationship between x and y across these subgroups. We also make panels of scatter plots when we want to compare the relationship between x and y across groups defined by a qualitative variable.

Example: *Relationship Between Price and Square Footage by Region*
Figure 4.10 is an example of a facet plot where each facet element shows the relationship between house price and square footage in one county. By separating scatterplots by county we can assess how the relationship between two quantitative values changes across counties. We can also see the range of prices and square footage that exist for each county. For example, we see that San Francisco only has data on small-area houses while other counties have more representation of larger houses. ∎

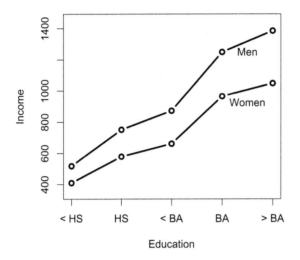

Figure 4.9 Income by Education and Sex. These line plots display the median weekly income in 2014 for full-time workers who are over 25. The dots for males (and females) are connected across education level, and for each education level, the dots for males and females appear one above the other.

Box-and-Whisker Plots for Groups We have seen that we can overlay density curves when comparing the shape of a distribution within subgroups. Side-by-side box plots is a simpler approach that can give a crude understanding of the relationship between one quantitative variable and two or more qualitative variables. When we have many subgroups, side-by-side box-and-whisker plots can be very useful for comparing basic features of a distribution.

Example: *Distribution of Housing Price by City*
Figure 4.11 shows box plots of housing price for various cities in the San Francisco East Bay. Each box plot represents the distribution of sale price (on a log scale) for houses in a city in 2004. The cities are ordered according to median sale price, which makes comparisons easier. ∎

Curse of Dimensionality Comparisons that involve more than one categorical variable can quickly become cumbersome as the number of possible combinations of categories grows. For example, there are $5 \times 4 = 20$ education–smoking status combinations, and to examine the distribution of birthweight across all 20 corresponding subgroups is difficult. Further, we come up against the problem of having too few observations in some subgroups. Although, there are more than 1,200 records in the

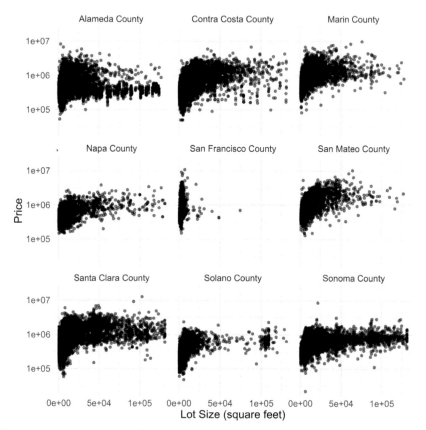

Figure 4.10 Housing Price and Square Footage by County. Each panel shows the relationship between square footage and price, separated by county. We can see how the relationship between these two quantitative values changes by region as well as see the range of prices and square footage that exist for each county.

CHDS, 7 of the 20 education–smoking combinations have fewer than 25 observations. See also Section 4.4.3 for other problems that occur when we have too few data. This "curse of dimensionality" also arises when we compare relationships with quantitative data. With just three variables, some of the plots in a panel of scatter plots can easily have too few observations to confirm the shape of the relationship between two variables across subgroups determined by a third.

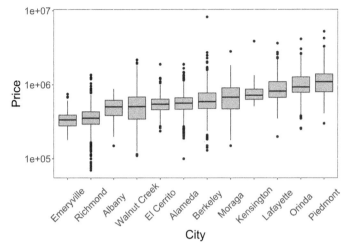

Figure 4.11 House Prices by City. This set of box plots display the summary statistics for sale price (log scale) of houses in 12 cities in the San Francisco East Bay in 2004. The cities are arranged according to their median sale price, which eases comparisons.

4.4 Examining the Visual Argument

Every plot has a message, and applying strategies for making this message stand out can make a big difference in the plot's readability and a reader's comprehension. The choice of scale, transformation, and order of bars are examples where the "right" choice can help a reader better understand the story. In this section, we collect advice on these aspects of visualization and organize them into five categories that address how to: incorporate the study design, choose scales, handle large amounts of data with smoothing and aggregation, facilitate meaningful comparisons, and add contextual information. By taking time to craft an effective plot, we help the reader quickly understand the plot's argument.

4.4.1 Incorporating the Data Design

When data have a geographic component, we should check for spatial patterns with, e.g., a map, or by conditioning on the geographic region when we make a graph (see the box plots of housing price by city in Figure 4.11). For data collected over a period of time, we should look for time trends by, e.g., making a line plot with time plotted along the x-axis (see Figure 4.12 and the accompanying example). If we want to remove or control time trends and spatial patterns, we can examine a subgroup defined by a particular time period or geographic region, e.g., the box plots in Figure 4.11 are for 2004 only.

Example: San Francisco Housing Price Deciles

Figure 4.12 shows weekly sale prices for homes in the San Francisco Bay Area from April 2003 through 2008. Housing prices fluctuated tremendously over this short time period as the housing market boomed and crashed. Plotted here are the deciles for weekly sale price; that is, each line tracks a decile (10th, 20th, 30th, ..., 90th percentile) of price for weekly sales. Each decile curve has been scaled to the first week's prices, and they are color-coded using a grey-scale that ranges from dark grey for the 10th percentile to light grey for the 90th percentile. We can see in this plot that the lower-priced homes suffered the greatest volatility in this period. Proportionally, they surpassed the higher-priced homes in 2006 and lost the greatest value in 2008. ∎

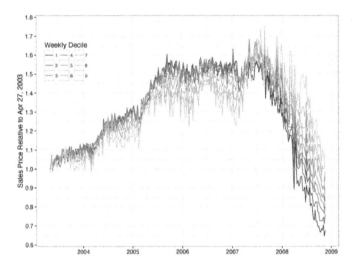

Figure 4.12 San Francisco Housing Prices. These nine line plots show the change in weekly housing prices over a six-year period. Each curve represents a decile in weekly house prices. These are scaled to their respective decile value for sales in the week of April 27, 2003 (the beginning of the data-collection period). The least-expensive houses show the greatest variability, rising the most and dropping the most, relative to their starting position.

In addition to geo-temporal variation, we should account for situations when individuals/observations have unequal chances of being selected for the sample. That is, we need to incorporate the sampling design into the plot. For example, if men have been sampled at twice the rate of women, then we must take this difference into account when we make, say, a histogram of income for the combined group of men and women. Otherwise, our plot might not accurately reflect the income distribution of individuals. We need to be particularly cautious with data that do not form a census or scientific sample. Within limitations, visual comparisons between subgroups can be effective in this situation (see Section 4.3).

We should also take care with cross-sectional data and not confuse patterns that we find with a possible cohort effect, as shown in the following example.

Example: *Race Times in the Cherry Blossom Ten-Mile Run*
The box plots on the left in Figure 4.13 show the race time (minutes) by age for male runners in the annual Cherry Blossom Ten-Mile Run from 1999 to 2012. It is tempting to interpret the curvature in the lower quartile, median, and upper quartile of the box plots as the trend in how an individual's performance changes with age. However, this is a cohort study, not a longitudinal study. The 25-year olds running in, say, 2001, are a different group of people than the 50-year olds running in the race that year, and these two groups could be different in ways that would affect the relationship between race time and age. For example, the 50-year olds in the run are likely to be more fit for their age than the 25-year olds.

Furthermore, these data involve a year cohort, i.e., the year of the run. The plot on the right conditions on the run year. Each curve is a local average of race time for runners of the same age in that year. The plot reveals interesting structure: average race times have slowed over the years, which is likely due to the increased popularity of the race and so higher participation from novice runners in the more recent years. ■

4.4.2 Choosing the Scale to Reveal Structure

The right choice of statistical graph can reveal features of the data such as the shape of a distribution and the relationship between variables (see Sections 4.2 and 4.3). However, there are also many second-order considerations in plot-making that can impact how easy it is for the reader to see important features in the data. We need to consider the scales of the axes and transformations of variables. Typically, the software we use to make a graph does a good job of choosing the scale, but not always. Sometimes to help the reader, we need to specify the scale and even the location of the tick marks on an axis when we create a plot.

Filling the Data Region It can be difficult to read a distribution when most of the data appear in a small portion of the plotting region. Features such as multiple modes and skewness can be hard to discern. Similarly, it can be hard to assess nonlinearity in a relationship when the points in a scatter plot do not fill the data region. Sometimes this phenomenon is due to a few unusually large observations. In order to get a better view of the main portion of the data we might drop these observations from the plot. If we do not include all of the data in the graph, then we must mention this exclusion in the caption or on the plot itself.

Example: *San Francisco Housing Price Distribution*
The density plot in the left-hand plot of Figure 4.14 displays the distribution of the sale price of all houses sold in 2004 in San Francisco. It is difficult to closely examine the distribution because a few unusually expensive houses force the bulk of the distribution into a small portion on the left side of the plot. In contrast, the density curve in the right-hand plot excludes these high priced houses. The shape of the distribution for the bulk

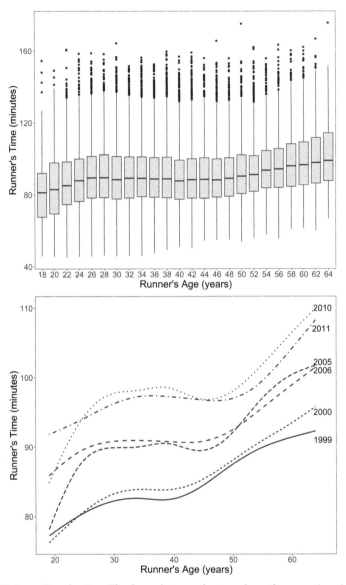

Figure 4.13 Race Time by Age. The box plots on the top show the race time (minutes) by age of the male runners in the Cherry Blossom Ten-Mile Run from 1999 to 2012. The plot on the bottom contains a smooth curve of race time for each of six years. One feature apparent from this plot is that average race times have slowed over the years the run was held.

of the houses is much clearer in this plot. There we can more easily observe the skewness of the main mode. ∎

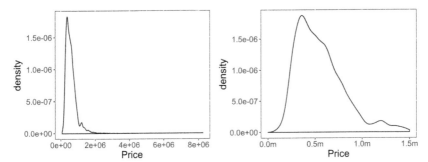

Figure 4.14 Distribution of Housing Price. Both plots show the distribution of sale price of houses in 10 cities in the San Francisco Bay Area in 2004. The one difference between them is that sales more than $1.5 million are excluded from the plot on the right. This exclusion makes it easier to see the shape of the bulk of the data. The distribution has a short left tail with a market entry point of about $250,000 and a mode around $350,000. The distribution is skewed right with a large right shoulder indicating that many houses are in the $500,000-$750,000 range. Even clipping houses that sold for more than $1.5 million, we see that the distribution has a long right tail with many houses selling for $1 million and more.

Including Zero Often there is no need to include 0 on an axis, especially if including it makes it difficult to fill the data region. In Figure 4.8, the y-axis begins at 58 oz because the smallest babies weigh more than that. On the other hand, the x-axis in Figure 4.3 includes 0 so that the lengths (positions of the dots) can be directly compared, i.e, we readily see that the proportion of mothers who smoked until pregnant is one-fourth the size of the current smokers.

Banking to Decipher Relationships With scatter plots, we try to choose a scale so that the relationship between the two variables roughly follows a 45-degree line (called banking to 45 degrees) because it makes it easier for the reader to see shape and trends. That is, it is much easier for us to determine departures from a line when the data roughly fall along a 45-degree line within the plotting region.

Revealing Shape Through Transformations Filling the data region is often a problem with highly skewed distributions, and a log transformation can help fill the data region. Additionally, when a transformation results in a symmetric distribution, this provides useful information about the distribution. For example, if a density curve of log-transformed values is approximately normal in shape, then we know the distribution of the original values is roughly log-normal.

Example: *San Francisco Housing Price Distribution, Continued*

In Figure 4.14, we clipped the highest-priced houses from the data to reveal the shape of the bulk of the distribution of prices. Alternatively, we examine the distribution of sale price on a log scale in Figure 4.15. We still see features that were evident in Figure 4.14, such as the shoulder on the left side of the mode and a possible smaller mode near $1.2 million. However, now we also see a roughly symmetric distribution. ■

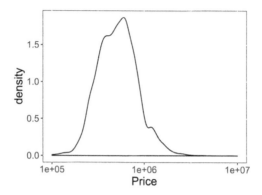

Figure 4.15 Distribution of Log-transformed Housing Price. This plot of the distribution of sale price of houses in 10 cities in the San Francisco Bay Area in 2004 is plotted on a log scale. The distribution of log price is somewhat symmetric, but the shoulder and smaller mode near $1.2 million are both still evident (see the right-hand plot in Figure 4.14 for comparison).

Revealing Relationships Through Straightening As described in Section 4.2.2, since it is easier for us to recognize linearity that to distinguish between different non-linear relationships, we apply transformations to create visual displays of linear relationships. We call this straightening, and it helps us assess the form of nonlinearity.

4.4.3 Aggregating and Smoothing Data

With large amounts of data, we often want to visualize aggregates of the data, rather than individual observations. For example, a dot plot marks each observation by a dot along the x-axis, and when we have more than a handful of observations, the dot plot makes a poor choice for visualizing the data when we have too many data points. Dots are plotted on top of one another (called "over-plotting") and there are too many plotting symbols to see the shape of the distribution. The scatter plot is the two-dimensional analog to the dot plot because each dot in a scatter plot represents an observation, i.e., a pair of values. Similar to the dot plot, when there are too many observations in a scatter plot, the relationship between the variables is obscured by over-plotting (see the scatter plot on the left in Figure 4.16).

Smooth to Uncover Shape Smoothing can address problems with over-plotting. Histograms and density curves smooth the location of observations and can reveal an underlying shape of the distribution. The histogram smooths observations across the bin in which they fall. We don't know the exact location of the observations in a bin and treat them as if they are evenly distributed across the bin, and a density curve provides an even-smoother representation of the distribution as it has no discontinuities.

Smooth to Uncover Trends With large amounts of data, we can replace individual values with aggregates. Specifically, we can plot an average value of y given a particular value of x (or narrow range of x values). The resulting curve can make it easier to read trends in relationships.

Example: *Cherry Blossom Ten-Mile Run, continued*
The scatter plot on the left in Figure 4.16 shows the race time (minutes) and age for male runners in the Cherry Blossom Ten-Mile Run from 1999 to 2012. There are more than 50,000 points in the scatter plot and over-plotting makes it impossible to see any relationship. In contrast, the plot on the right smooths the data by taking local averages of the race times for runners of the same age in each year of the run. As noted earlier, this plot reveals that average race times have slowed over the years. ∎

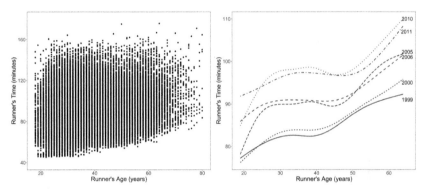

Figure 4.16 Race Time by Age. The scatter plot on the left shows the race time (minutes) and age for male runners in the Cherry Blossom 10-mile run from 1999 to 2012. There are over 50,000 points in the scatter plot. Due to over-plotting we cannot see any patterns in the data. The plot on the right contains a smooth curve for each of six years of the run. One feature apparent from these curves is that race times have slowed over the years.

When Not to Smooth When we have only a handful of observations, smoothing techniques can give misleading representations of the data. With just a few observations, we prefer rug plots over histograms, box plots, and density curves, and we make scatter plots rather than smooth curves and density contours.

Reducing Distributions to Quantiles The normal-probability plot compares the quantiles of the data to those of the theoretical normal curve. This plot is essentially a scatter plot of the quantile pairs: (qth quantile of the normal distribution, qth quantile of the data). If the data's distribution is close to normal, then the points roughly fall on a line. Deviations from a line tend to be easier to read with quantile plots than with overlaid histograms or density curves. We can also compare the quantiles of two sets of data. Figure 4.17 plots the quantiles of birthweight for never and current smokers. There we see that the points roughly fall on a line, which indicates the distributions have roughly the same shape. However, we also see that the line is shifted down from the reference line of intercept 0 and slope 1, which indicates a shift left in the distribution of weight for the babies born to smokers in comparison to never smokers.

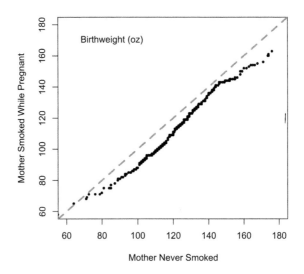

Figure 4.17 Quantile-Quantile Plot of Birthweight. This plot compares the quantiles of birthweight for mothers who smoked during pregnancy to those who never smoked. The quantile pairs roughly fall on a line which indicates the two distributions have similar shapes. The dashed line added to this plot has slope 1 and intercept 0. The downward shift of the points compared to the reference line indicates lower birthweights for smokers.

Reducing Shape to a Box-and-Whisker Plot Summary statistics, such as the median and interquartile range (IQR), give us an idea of the center and spread of a distribution, and the box-and-whisker plot offers a visual summary of these statistics. Box-and-whisker plots also can indicate symmetry, but they are not as informative as a density curve or histogram because they cannot, for example, show modes. Box-and-whisker plots (also known simply as box plots) tend to be most useful when comparing summary statistics across many groups. See for example Figure 4.11.

Revealing Density in a Scatter Plot We can alleviate problems with over-plotting in a scatter plot in a variety of ways. We can make the color of the plotting symbols partially transparent so that when plotting symbols overlap the color accumulates and appears darker. Alternatively, we can make a contour plot to reveal the density of points. If many observations have the same values (e.g., due to recording values to the nearest inch), then we can add a small amount of random noise to the values to reduce the amount of over-plotting. This procedure is called "jittering". For example, the heights of the mothers and fathers in the CHDS were measured to the last completed inch, and to avoid over-plotting, we have added random noise to both parents' heights in Figure 4.4.

4.4.4 Facilitate Meaningful Comparisons

When we create a plot, we keep in mind the important comparison we want our plot to demonstrate so we construct the plot to make it easier for the reader to discern this comparison. We might for example, superimpose two density curves, or juxtapose a grid of scatter plots. We have seen already in Figure 4.17 that when we want to compare the distribution of values for two groups that the quantile-quantile plot is a useful statistical graph. We have also seen in Figure 4.11 that we can split our data into subgroups, according to the value of a third variable, and compare the resulting conditional distributions with specially ordered side-by-side box plots.

Emphasize the Important Difference When we examine relationships between qualitative variables, we examine proportions of one variable within subgroups defined by other variable(s). We determine the proportions to compute based on what we consider to be the important comparison. Similarly, when we compare averages of a quantitative variable across groups defined by two or more qualitative variables, we have choices for how to plot these averages. We want to emphasize the important differences, as in the example below.

Example: Weekly Earnings for Men and Women in 2014, Continued
When we examine the average income across sex and education, the comparison of interest is between men and women at each education level. The two plots on the top in Figure 4.18 show two arrangements. In the plot on the right it is easier to see that at all education levels, women earn less than men on average. Additionally, we see how the gap between men and women increases with education. It is difficult to compare the wages of men and women in the plot on the left because our eyes must travel across half the plot to compare the wages of men and women at the same education level. The left plot emphasizes the less-interesting difference in income across education. ■

Order by Proportion The main advantage the bar plot has over a table of statistics is in the ease of making relative comparisons. To further assist in comparing bars, we

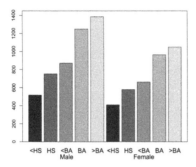
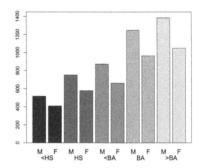

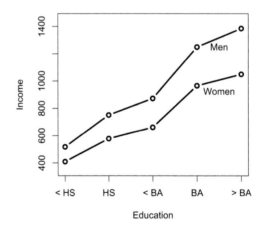

Figure 4.18 Income by Education and Sex. These plots compare the median weekly income in 2014 of full-time workers over 25. In the bar chart on the top left, the bars are arranged by education, separately for males and females. This arrangement makes it difficult to assess the more interesting differences between men and women with the same education level. The bar chart on the top right assists the reader in making the comparison between men and women, and the line plot on the right facilitates this comparison even more because the dots for males and females appear one above the other at each education level. With the bottom plot, we can more easily see how the gap in income between men and women grows with education.

order them according to their height (as in Figure 4.3), unless the natural ordering of the categories has meaning, e.g., education level. When we compare two or more qualitative variables, we normalize group counts and use proportions.

Order by Median With side-by-side box plots we arrange them in order according to their medians (as in Figure 4.11), unless, again, the variable has a natural ordering.

Avoid Stacking Stacked bar and line plots hinder comparisons as they lack a constant base line. With the exception of the lowest portions of each bar, the bottom of a sub-rectangle in a stacked bar plot moves up and down across groups. Stacked line plots are even more difficult to read as the distance from one curve to another wiggles up and down making trends hard to read.

Example: CO$_2$ *Emissions*
Both plots in Figure 4.19 show CO$_2$ emissions from 1950 to 2012 for the 14 countries with the highest emissions. In the left plot, the emissions for a country at a point in time is conveyed by the height of the corresponding color band. It is very difficult to read the shape of the growth in emissions for a country because the color band wiggles up and down. In the right plot, the time series for each country is plotted relative to the x-axis, i.e., the countries all have the same base (the x-axis), and the base doesn't move up and down with year. Now the trends over time are much clearer. We see that some countries have had flat rates of growth in CO$_2$ emissions (on a log scale), and others have increased much faster. Countries are more easily ranked by annual emissions. ∎

Compare Lengths Rather than Angles A pie chart conveys proportions through angles in the pieces of the pie. Comparisons between slices in a pie can be difficult to make, in contrast to lengths of bars or the locations of dots along a line. For this reason bar charts and dot charts are typically preferred to pie charts.

Compare Points Along a Line Although line plots and bar plots are similar in how they convey information, when subgroups are to be compared, the line plot has advantages over the bar chart. As an example consider the bottom plot in Figure 4.18. The dots for men and women within each education level appear directly above/below one another, and the dots for the same sex are connected across education levels. These features make it easier for us to make comparisons. The growth in the gap between men and women with increasing education is clear in the line plot and harder to discern with the side-by-side bars.

Decide Whether to Superimpose or Juxtapose When we juxtapose plots, i.e., make a panel of plots, we typically use the same limits on the axes of all the plots in order to facilitate comparisons across plots (see the facet plot in Figure 4.10). However, depending on the number of comparisons and the amount of data, it can be easier to superimpose plots (place them on top of each other) so our eyes don't have to move back

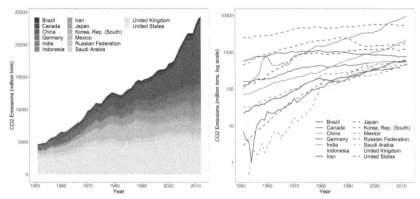

Figure 4.19 CO_2 Emissions by Country Over Time. These plots show the growth in CO_2 emissions (million tons) from 1950 to 2012 for various countries. The main difference between the two plots is that the left plot stacks one country's data on another's so the value of CO_2 for a country is the vertical difference between two curves. On the other hand, the plot on the right superimposes each country's data in a time series, which makes it much easier to examine the trends for individual countries. For example, we can see easily that some countries have roughly flat curves, which indicate exponential growth (the y-axis is on log scale), other countries are increasing, and a couple are decreasing (e.g., Germany). However, on the right, we are forced to use color to help distinguish between different countries' lines.

and forth between plots. See for example the superimposed density curves in Figure 4.6 and the line plots in the bottom of Figure 4.18.

Select a Color Palette We can use color to add another variable to a plot, e.g., we can change the colors of the dots in a scatter plot to represent the values of a qualitative variable. We also can use color to represent magnitude, such as the size of a proportion. The choice of colors depends on the kind of information conveyed. For categories of a qualitative variable, we use a collection of colors that are easily distinguishable and where any one color does not stand out more than another. For quantitative data, we use a sequential gradation that emphasizes one end of the spectrum of values over the other, or we employ a diverging palette that emphasizes both extremes and not the middle. For example, with cancer rates, we want to use a sequential palette that runs from light to dark for higher rates. On the other hand, with two-party election results, a diverging palette with two distinct hues for low and high values distinguishes between the dominance of each party. More about the use of color appears later in this section.

4.4.5 Adding Contextual Information

We want our statistical graphs to be rich with information that does not distract from the main message. We can add information to a statistical graph with the use of plotting symbols, colors, reference markers, and point labels. This additional information provides a context for interpreting the findings. Every element of a statistical graph should have a purpose. Superfluous text or plot features, often referred to as "chartjunk", should be eliminated. Additionally, we want to clearly describe our insights in a comprehensive caption.

Dimensions and Symbols We can add information to a statistical graph with the use of color and different plotting symbols (see Figure 4.8). Each dimension of a multidimensional plot (e.g., the depth of a three-dimensional bar) should represent a measured quantity, otherwise it is unnecessary and should be excluded from the plot.

Color The use of color should not be gratuitous; that is, colors should represent information. Plots are meant to be examined for long periods of time so we choose colors that do not impede the reader's ability to take time to carefully read a plot. We want to avoid colors that create an after-image when we look from one part of the graph to another, colors of different intensity that make one attribute appear more important than another, and colors that color-blind people have trouble distinguishing between. Furthermore, people typically have trouble distinguishing between more than about seven colors so we limit the number of colors used in a plot. Finally, colors can appear quite different when, say, printed on paper or viewed on a computer screen so when we choose colors, we keep in mind the medium in which our plots will be displayed.

Reference Markers Reference points and lines can provide benchmarks, historical values, and other external information to compare our data against and help interpret the results. For example, we can compare the relationship between two quantitative variables to a line with slope 1 by simply adding this reference line to a scatter plot (see Figure 4.17). Or, we can add a vertical line to a density curve to see where the distribution falls in relation to an important benchmark.

Summaries When we have a lot of data, we often want to de-emphasize the individual values and focus on the distribution, relationships, and trends. To do this, we smooth, average, and augment plots with additional summary statistics (see Section 4.4.3 for examples).

Labels It is good practice to consistently use informative labels on tick marks and axes (including units of measurement). Our graphs should contain titles, and when needed, legends. If a plot is part of an informal exploratory analysis, we can include less contextual information than for a formal statistical graph. However, even with exploratory data analysis (EDA), we want to include enough context so that when we return to an analysis we can easily figure out what we plotted. Note that the titles of plots sometimes appear in the figure captions as they do in this chapter.

Captions Plot captions serve several purposes. They describe what has been plotted and orient the reader. Captions also point out the important features of the plot and their implications. It is okay for the caption to repeat information found in the text. Readers often skim an article and focus on section headings and visualizations so plot captions should be self-contained. More information about writing captions can be found in Section 7.3.5.

4.5 Editing and Revising

Making good statistical graphs is hard. It usually takes an iterative approach where we progressively improve our plot with a change of scale, transformation of a variable, addition of information, color, and reference markers, etc. We may even decide it is best to entirely discard our plot and start over with a new approach all together. This is not time wasted because each of these steps engages us in better understanding our data and models. A typical sequence to create a plot goes as follows:

- Select a plot type according to the kind of data.

- Make the plot taking the software defaults and using cryptic labels.

- Consider transformations to symmetrize and straighten relationships and the choice of scale.

- Address any issues with over-plotting.

- Consider variable(s) to condition on that would inform whether the shape or relationship observed is maintained across subgroups.

- Try different approaches to visualizing this comparison, such as a grid of plots, use of color and plotting symbols, and smoothing, etc.

- Determine whether there is any additional information that would help put the findings in context, such as reference markers or highlighting particularly interesting observations or features.

- Add informative labels, legends, and titles.

- Write the caption.

This process is similar to the writing process. The default statistical graph produced by the software is like a first draft, and we put a large effort into revising to improve the story we are telling. On occasion we even decide to discard our current draft and begin again. In the moment, we might think some of the steps listed above are unnecessary for a particular plot, but it is usually a good idea to work through each step. As we go from one step to the next, we might uncover something that we didn't expect to find. This may signal a need to redo an earlier step. We provide an example of a cautionary tale about not getting caught up in the editing process before finding the best plot for telling our story.

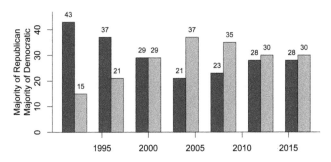

Figure 4.20 Bar Chart of Majority Party in California Counties. The side-by-side bars show the number of counties in California that have a majority of voters registered Republican (pale gray) or Democrat (dark gray). Each pair of columns adds to 58, the number of counties in California. This plot was made on swivel.com (no longer a live site) from data available at the state of California's voter registration site.

Example: California Voter Registration Trends
We are interested in voter trends in California and visit the online voter registration site of the California Secretary of State. There we find county summaries of voter registration for seven presidential election years (1992, 1996, ..., 2016). We scrape these data and make the plot in Figure 4.20.

This plot has many problems, as listed below.

- Tick marks: x-axis tick marks are at five-year intervals and do not line up with the locations of the bars so the reader has to work too hard to figure out that the bars correspond to measurements made at four-year intervals.

- Color: atypical use of light and dark green for the Democratic and Republican parties, which are traditionally represented by blue and red.

- Legend: lack of a legend means that we cannot discern which color represents which party.

- Axis label: y-axis label does not indicate the units of measurement.

- Title: confusing title that does not illuminate the content of the plot.

It is tempting to quickly try to fix these problems. We can "improve" the plot (see Figure 4.21), but we should first think some more about the story and whether this plot makes the best argument. Our aim is to show change in voter registration over the past seven presidential elections. However, it is people who register to vote, not counties. County size is a lurking variable—small counties tend to be rural and conservative—so using counties overstates the Republican presence. Rather than record counties, we should tally voter registration counts. To do this, we revisit the registration website to obtain new data. There we find a page of voter registration counts by party, including counts for other parties and for unaffiliated voters. Since the number of unaffiliated voters is sizable, it could be informative to include these registration numbers in the plot.

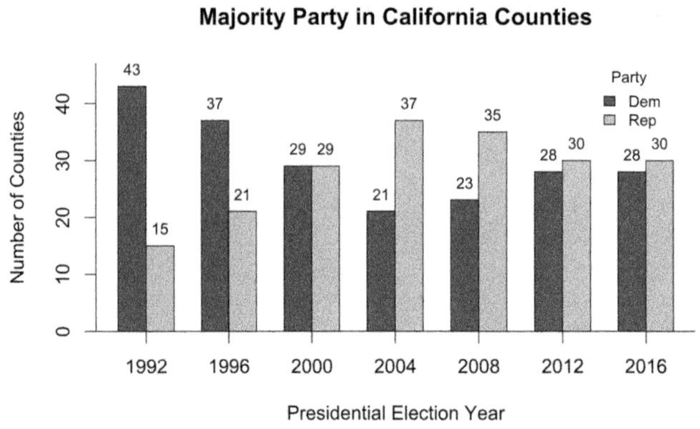

Figure 4.21 Bar Chart of Majority Party in California Counties—Revised. This bar chart addresses many of the problems found with the bar chart in Figure 4.20, including the inaccurate y-axis label, ill-positioned tick marks on the x-axis, and lack of legend.

What kind of plot should we make? We have registration figures over time so a line plot seems appropriate. Also, given that the California population has grown dramatically in the past 25 years, rather than compare raw registration numbers, we scale them by each year's total registration and compare percentages.

We discard the first plot and make an entirely new one (see Figure 4.22). This new graph makes a more interesting and accurate depiction of voter registration trends in California. We see that: the percentages of registered Democrats and Republicans have declined over this period; the percentage of Democrats was about 10% higher than the Republicans in the earlier years but the spread has grown to about 15%; and the percentage of unaffiliated voters has dramatically increased from about 10% to 25%.

The time spent editing, revising, and entirely remaking our graph was well spent. We now have a plot that makes a compelling visual presentation of the change in California voter behavior over the last 25 years. ∎

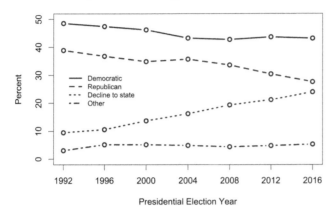

Figure 4.22 Distribution of California Voters by Party. The line chart addresses the essential problem with the bar chart in Figure 4.21, i.e., we are interested in the change in voter registration over the years, not in the number of counties that are majority Republican or Democratic. Here we see that the percentage of registered Democrats and Republicans have declined in this period, the percentage of unaffiliated voters has dramatically increased, and the gap between Democrats and Republicans has grown from about 10% to 15%.

4.6 Notes

In this chapter, our focus is on how to "read" and "write' statistical graphs. There are several excellent sources for learning more about data visualization, including books by Cleveland [5], Tufte [10], and Wilkenson [12].

The code and data to make this chapter's plots are available on the book's website. Many of the plots in this chapter use the CHDS data described in Section 3.5. Figures 4.11 and 4.12 use data on housing sales in the San Francisco Bay Area from 2003 to 2008, scraped from the *San Francisco Chronicle* site of weekly sales (no longer available). Figure 4.12 is modeled after a figure in Wickham, Swayne, and Poole's chapter in *Beautiful Data: The Stories Behind Elegant Data Solutions* [11]. Figures 4.13 and 4.16 use data on run times scraped from the Credit Union Cherry Blossom site [6]. The data for the visualization about earnings of men and women (Figures 4.9 and 4.18) are from the 2014 Current Population Survey on the Bureau of Labor Statistics [3]. Figure 4.19 on CO_2 emissions visualizes data provided by the World Resources Institute via their CAIT Climate Data Explorer [13].

We first came across the visualization on California voters (Figure 4.20) at swivel.com, which is no longer an active site. The data for Figures 4.20, 4.21, and 4.22 are from the California Secretary of State site [4].

Cynthia Brewer's palettes are handy for choosing sequential, diverging, and qualitative colors. These palettes are available in Python and R, and her online app provides an interface for choosing colors [2].

More activities for data visualization appear in Nolan and Perret [9] and Chapter 4 of Gelman and Nolan [7].

For more practice, consider working on Portfolio Pieces I, M, N, R, or T.

4.7 References

[1] J. Bartlett. Running simulation studies in R—an introductory tutorial. https://htmlpreview.github.io/?https://github.com/jwb133/SimStudiesR/blob/master/SimulationStudiesinR.html#a-very-simple-simulation-study, April 2019.

[2] C. Brewer and M. Harrower. COLOR BREWER 2.0: color advice for cartography. http://colorbrewer2.org/.

[3] Bureau of Labor Statistics. Labor force statistics from the Current Population Survey. https://www.bls.gov/cps/earnings.htm, November 2019.

[4] California Secretary of State. Voter registration statistics, 2020. https://www.sos.ca.gov/elections/voter-registration/voter-registration-statistics/.

[5] W. S. Cleveland. *Visualizing Data*. Hobart Press, 1993.

[6] Credit Union Cherry Blossom. Credit Union Cherry Blossom Ten Mile Run & 5K Run-Walk, 2019. http://cherryblossom.org/.

[7] A. Gelman and D. Nolan. *Teaching Statistics: A Bag of Tricks*. Oxford University Press, Oxford, 2nd edition, 2017.

[8] T. McCann. Monte Carlo models in Python. https://www.advancinganalytics.co.uk/blog/2018/12/12/monte-carlo-models-in-python, December 2018.

[9] D. Nolan and J. Perrett. Teaching and learning data visualization: Ideas and assignments. *The American Statistician*, 70(3):260–269, 2015.

[10] E. R. Tufte. *The Visual Display of Quantitative Information*. Graphics Press LLC, 2001.

[11] H. Wickham, D. F. Swayne, and D. Poole. Bay Area blues: The effect of the housing crisis. In T. Segaran and J. Hammerbacher, editors, *Beautiful Data: The Stories Behind Elegant Data Solutions*. O'Reilly Media, Sebastopol, CA, 2009.

[12] L. Wilkenson. *The Grammar of Graphics*. Springer, New York, 1999.

[13] World Resources Institute. CAIT climate data explorer. http://cait.wri.org/.

4.8 Activities

4.8.1 Deconstruct and Reconstruct a Plot

The process of identifying and deconstructing a problematic graph and then reconstructing it in a more appropriate form can help us better interpret, critique, and construct meaningful graphs. In this activity, you are to first critique the appearance and application of a statistical graph. Then, create your own graph by following the guiding principles of Section 4.4. More detailed instructions follow.

Ideally, this activity is done (and is more fun) in pairs, but you may do the activity on your own.

Part 1: Choose Your Plot Find a plot online and take a screenshot of it. The plot should have the following characteristics:

- The topic interests you and your partner;
- You can improve the plot;
- The data are available and complex enough to create an interesting graphic (look for two variables or more).

Part 2: Deconstruct Critique the plot.

- Determine what message you think the creator of the plot is trying to convey.
- Find at least two significant problems with the plot.
- Write a caption for this plot that describes the message and the problems that you found with it.

Part 3: Reconstruct Make a new plot.

- Fix the problems described in your caption for the original plot. You are free to make an entirely different type of plot that better conveys the story. Note that you may need to reshape the data or obtain more appropriate data for your revision.
- Add more information to the plot that will help convey the message more clearly or in a larger context. This additional information may be reference markers or additional data.
- Write a detailed caption for your version of the plot (a long paragraph). Describe what you have graphed. Point out the main features. State the message to be gleaned from the plot. Include references for any additional data or information used.

4.8.2 Finding a One-Minute Revelation

The goal of the one-minute revelation is to make a plot that shows a central finding that is part of a larger data-analysis project. This process can help organize your workflow as you dig deeper into a dataset (see also Chapter 6 on storyboarding with visualizations).

Your plot is expected to obey the principles of good graphics, but it is not meant to be a formal presentation graphic. For example, you are not expected to use additional information to make the plot information rich at this stage. Instead, focus on uncovering important features in the data. Consider working with others to synthesize a variety of findings from one dataset into a coherent story and present a plan for further investigation.

Create one graphical summary of your data and write a one-paragraph caption that describes your finding. If you are working with a partner or a group, coordinate with others so that the graphs tell a different part of the story.

These exploratory visualizations should:

- uncover important structure in the data,

- be interpreted in the context of the problem/question you are investigating, and

- set the stage for a formal analysis.

Write a one- or two- paragraph explanation of the direction your analysis would take in the more formal analysis. Tie this explanation to your findings. If working on a team with multiple figures, your explanation should refer to all of the figures when providing reasons for the more formal analysis plan.

4.8.3 Turning the Table on a Simulation Study

Many reports and articles display data summaries as tables rather than graphs, even though graphs are typically better suited for perceiving trends and making comparisons and predictions. To combat this convention, in this activity you will turn a table of results from a simulation study into a statistical graph.

First, simulate the large sample behavior of an estimator (choose something other than the mean, such as the trimmed mean) where you can vary two parameters (such as the amount of trimming and the mean of the data's sampling distribution). Run at least a few thousand simulations for different combinations of the parameters. Create a table of summary statistics from the simulation.

If you need guidance to get started on a simulation study, follow the approach in Bartlett's blog [1] (if you are an R user) or the approach in McCann's blog [8] (if you are a Python user).

Translate aspects of the table you made into visual elements and choose an appropriate graphical representation. Use the principles of the deconstruct-reconstruct activity (Section 4.8.1) to help you think about the story behind the simulation results and how to display the information in a compelling way.

Be sure to consider how to convey information about the variability in your simulation results. You may have to reorganize the data from the form used to make the table. What insights can you gain from the plot that were not readily apparent in the table? Write a caption that describes what you have plotted and the salient features.

4.8.4 Review of a Statistical Graph

In this activity you will exchange a plot, caption, and any code and data needed to produce the plot with a peer to get feedback on your visualization choices.

When you first receive your partner's plot, cover the caption. Examine the plot and write your own caption, including a description of the plot's main take-away. Then, compare your caption with your partner's. Did you interpret the plot correctly?

Here is a list of questions to consider to help you critique your partner's work.

- What are the variables displayed in the plot and how they are indicated (color, shape, etc.)?

- Does the plot type match the data types?

- Which features of the distribution, relationship, or trend stand out?

- Are transformations used? If so, are they effective?

- Is the scaling appropriate?

- Have the data been aggregated or smoothed? If so, is the approach appropriate?

- Are meaningful comparisons facilitated?

- Which of your partner's choices make it easy/hard to see the take-away message?

- Is there a way to simplify the graph while maintaining the message?

- Is any key information missing?

Run your partner's code to ensure that you can recreate the graphic. Make adjustments to the code based on your answers above. Return your version of the plot and caption, your updated code, and your other feedback to your partner.

5 Communicating Through Code

C ode serves as both a way to carry out a data analysis or a simulation study and a way to communicate our ideas about the analysis. Writing clear and simple code makes it easier for others to understand our thought process, check our analysis, or apply our methods to new data. Code that is easy to follow helps us too. It helps us prevent mistakes and remind us what we did when we need to explain or modify our code in the future. Just as we read and examined articles and reports that summarize data analyses to learn how others organize and write about their findings, we can learn a great deal about how to write code from reading code written by others.

This chapter discusses how to write pseudocode to give a high-level explanation of your approach, use technology tools to make sharing your work easier, and adopt coding conventions and principles so that your code is readable and easily understood by others.

5.1 Pseudocode

Pseudocode provides an overview of a computational task without the precise implementation details and without adhering to a particular programming language syntax. Pseudocode can be included in a publication to explain the implementation of an algorithm, an idea behind a statistical method, or the data-processing pipeline for an analysis; it keeps the focus at a high level and does not take up too much space. The full code used to produce the results is often provided online as supplementary material, rather than in the article. Ideally, the code is provided via a notebook that integrates explanatory text with code (see Section 5.5.2).

Pseudocode focuses on logic, not syntax, and can help us organize our thoughts and write better code. The fundamental elements in pseudocode are inputs, outputs, and key verbs that describe operations. We want to use concrete, specific, and active verbs

Communicating with Data: The Art of Writing for Data Science. Deborah Nolan and Sara Stoudt, Oxford University Press (2021). © Deborah Nolan and Sara Stoudt.
DOI: 10.1093/oso/9780198862741.003.0005

to concisely represent core actions. Although we do not need to specify the format of our inputs and outputs (e.g., data frame, list, dictionary, array, etc.), it can be helpful to consider what type they are (e.g., string, integer, double, factor, etc.). As we describe how to take the inputs and create the outputs, we need to consider the order of operations, including conditional and repeated evaluation of expressions, and ensure that a piece of information is presented ahead of a step that relies on it.

Example: *Bootstrap Pseudocode*
The pseudocode below explains how to compute a bootstrap estimate of the standard error of a sample median.

```
[1] Given data X = (X(1), ... , X(n)) and
[2] B = number of bootstrap replicates:

[3] T = median(X)
[4] Tboot = vector of length B

[5] Do B times, i = 1 to B
[6]    X* = sample of size n from X (with replacement)
[7]    Tboot(i) = median(X*)
[8] End

[9] SEhat = SD(Tboot)
```

The pseudocode first defines the input variables: the data, X; the number of observations, n; and the number of bootstrap replicates, B. The plain English descriptions of the computations give us the basic concept of the bootstrap procedure. Notice how the indentation helps clarify which part of the task is repeated. For operations, such as taking the median and standard deviation (SD), we expect the software to provide built-in functions for them so the pseudocode gives only names for these operations. An *R* implementation of lines 5–8 might be:

```
Tboot =  replicate(B, median(sample(X, n, replace=TRUE)))
se = sqrt(var(Tboot))
```

Here, the replicate() function handles the looping for us.   ∎
   Just as in writing text, there is a balance between being precise and concise when writing pseudocode. We need to include enough detail so that a reader can understand and possibly reproduce our approach, but not so much detail that the reader loses sight of the key computational features. If we use pseudocode as an outline before writing the actual code, it can help us write clearer, more focused code. The pseudocode can also act as documentation for our code (see Section 5.2.2).

5.2 Style Guidelines

Code has two "audiences"—computers and humans. To improve readability for the the human audience, we adopt a set of conventions for writing code, and we keep these conventions consistent within a project. Code that is easy to read is also easier to understand and modify. When we write prose, we also adopt conventions, e.g., for paragraph breaks, section titles, quotes, references, etc. The conventions we use may vary from one piece of writing to another, but they remain consistent throughout the piece. The same holds when we write code.

We recommend the following conventions and guidelines for naming functions and variables, formatting code so the control structure is visually clear, and providing helpful comments on what the code does.

5.2.1 Naming Conventions

We recommend using descriptive variable and function names. Variable names should suggest the nature of the values being stored, and function names should convey the purpose of the function. A good practice is to use nouns for variable names and active verbs for function names. The noun describes the kind of information in the variable, e.g., it can be helpful to adopt a convention for logical variables that makes it clear the contents indicate a condition, such as is_over_weight. Verbs should describe what the function does, e.g., the function create_DF() converts an object (e.g., a matrix of strings) into a data frame. These naming conventions reduce the amount of effort to read and understand code, and makes it easier to spot errors. If the names are well chosen, they provide additional information about the purpose of the code and make it easier to update and reuse code.

There are trade-offs to using short versus long names. Short names make for fast and easy typing, but if they are too short, it can be hard to keep track of what they stand for and avoid duplication. On the other hand, while longer names provide more information about a variable's contents, they can get in the way when we read code. A good compromise is to use short names for local variables and longer names for important values. However, we avoid short names that are overly simplistic and uninformative, such as x2, x3, temp, and foo. Also, if a name consists of two (or more) words, then we recommend either capitalizing the second (and later) words, e.g., isOverWt, or placing underscores or periods between words, e.g., is_over_wt. Whichever convention you choose, be consistent.

5.2.2 Comments and Documentation

No matter how concisely and clearly we have written our code, adding comments further clarifies the intent of our code. We recommend you comment code as you write it, and write pseudocode first, which then can be used as comments (see Section 5.1). Comments written at a high level aid the reader who is less familiar with the computational task so the reader can focus on the problem to be solved, rather than the particulars of your approach.

We suggest writing block comments that span multiple lines and pseudocode at the beginning of a function that briefly describes its purpose and explains the expected inputs and return value. Such documentation can help us determine whether the code accomplishes the intended task. Good documentation does not simply repeat the code.

When we write comments, we follow the guidelines in Kernighan and Pike's *The Practice of Programming*, especially their do's and don'ts:

- Do clarify; don't confuse.

- Do be succinct; don't belabor the obvious.

- Don't comment bad code; instead, rewrite the code.

- Don't contradict the code in the comments; keep comments and code in sync as you rewrite your code.

Example: *Google's R and Python Guidelines*
According to Google's guidelines for coding in *R*,

Functions should contain a comments section immediately below the function definition line. These comments should consist of a one-sentence description of the function; a list of the function's arguments, denoted by Args:, with a description of each (including the data type); and a description of the return value, denoted by Returns:. The comments should be descriptive enough that a caller can use the function without reading any of the function's code.

These guidelines include the following example of documentation for the Calculate-SampleCovariance() function.

```
CalculateSampleCovariance <- function(x, y, verbose=TRUE) {
  # Computes the sample covariance between two vectors.
  #
  # Args:
  #   x: 1 of 2 vectors whose sample covariance is calculated.
  #   y: The other vector.
  #      x and y must have the same length,
  #      greater than one, with no missing values.
  #   verbose: If TRUE, prints sample covariance; if not, not.
  #      Default is TRUE.
  #
  # Returns:
  #   The sample covariance between x and y.
  ...
```

Notice that the function name uses camel case (capitalization of second and later words), and describes with an active verb what the function does.

According to Google's guidelines for coding in Python, a function is documented via a docstring.

A docstring should give enough information to write a call to the function without reading the function's code. The docstring should be descriptive-style ("""Fetches rows from a Bigtable.""") rather than imperative-style ("""Fetch rows from a Bigtable."""), except for @property data descriptors, which should use the same style as attributes. A docstring should describe the function's calling syntax and its semantics, not its implementation. For tricky code, comments alongside the code are more appropriate than using docstrings.

These guidelines include the following example of documentation for the function fetch_bigtable_rows().

```python
def fetch_bigtable_rows(big_table, keys, other_silly_variable=None):
    """Fetches rows from a Bigtable.

    Retrieves rows pertaining to the given keys from the Table
    instance represented by big_table.  Silly things may happen
    if other_silly_variable is not None.

    Args:
        big_table: An open Bigtable Table instance.
        keys: A sequence of strings representing the key of each
            table row to fetch.
        other_silly_variable: Another optional variable, that has a
            much longer name than the other args, and which does
            nothing.

    Returns:
        A dict mapping keys to the corresponding table row data
        fetched. Each row is represented as a tuple of strings.
        For example:

        {'Serak': ('Rigel VII', 'Preparer'),
         'Zim': ('Irk', 'Invader'),
         'Lrrr': ('Omicron Persei 8', 'Emperor')}

        If a key from the keys argument is missing from the
        dictionary, then that row was not found in the table.
```

```
Raises:
    IOError: An error occurred accessing bigtable.Table object.
"""
```

Here, naming uses underscores, but the description of the function is still active. ∎

5.2.3 Whitespace in Expressions

Although often not required, using whitespace in expressions can greatly aid readability. As a first step, we recommend putting a single whitespace or blank after a comma, before and after infix operators (such as + and =), and before a left parenthesis (except in a function call). Placing blank lines around comments and paragraphs of code also aids in readability.

For more whitespace guidance we recommend following the Python Enhancement Proposal (PEP 8) guidelines. These guidelines work for programming in *R* as well. Specifically, they suggest always surrounding binary operators with a single space on either side, but to not use spaces around the = when it indicates a keyword argument or default parameter value. The recommendations include guidelines for avoiding extraneous whitespace, such as immediately inside parentheses, brackets and braces, and before the open parenthesis of a function call. They also suggest not using multiple whitespace to align assignment statements with one another.

5.2.4 Control Structure and Indentation

Some languages require particular indentation to be syntactically correct. Even when a language does not require this, such as with *R*, we strongly recommend using consistent indentation to denote the structure of the code. We generally align code that is at the same level and indent code with two spaces for each sub-level. For example, expressions within a function are indented two spaces; if there is a loop in the function, then the code within the loop is indented an additional two spaces; and if there is a conditional statement nested within this loop, then the code block in this construct is indented another two spaces. Other conventions use more spaces, e.g., four spaces or one tab. No matter your preference, consistency helps to easily identify control flow.

We also recommend using parentheses when they are not strictly required in order to resolve ambiguity. In *R*, we adopt the convention of placing the closing curly brace on its own line, with the exception of an `else` construct where we always put the `else` on the same line as the closing curly brace of the `if` code block, i.e.,

```
if (condition) {
  statements
} else {
  statements
}
```

Additionally, we do not place opening curly braces on their own lines, with the exception of the opening curly brace for a function definition. Curly braces can be omitted when the function or code block contain only one expression. We recommend choosing one of these conventions (use curly braces or not with a one-line code block) and apply it consistently.

Example: *Function Formatting*

Compare the way the following two R functions are formatted. Both are syntactically correct. However, the first version uses no indentation or extra whitespace.

```
function(x,epsilon=NULL,non_neg=FALSE) {
if(non_neg){if(!is.null(epsilon)){ x[x<=0]=epsilon
}else warning("no epsilon value provided")}
mean(x,trim=0.1)}
```

The second version indents two spaces for each level of control and has spaces after commas and before and after operators.

```
function(x, epsilon=NULL, non_neg=FALSE)
{
  if (non_neg) {
    if (!is.null(epsilon)) {
      x[x <= 0] = epsilon
    } else warning("no epsilon value provided")
  }

  mean(x, trim=0.1)
}
```

The indentation makes the nesting of the `if` statements clear. The spaces surrounding operators, such as `x <= 0`, and following commas, such as `mean(x, trim=0.1)`, also help to highlight the code structure.

Note that in Python line spacing and indentation matter for interpretation and execution of code. ∎

5.2.5 Line Length

Long lines of code can be hard to read, especially if we have to scroll in order to see the entire expression. We suggest breaking long expressions across multiple lines so that we can see the entire expression within a window of 60–80 characters. To do this, we typically break function calls at commas and align the new line of code with the first argument in the previous line. Fortunately, many code editors automatically perform good indentation practices. Although permissible in some languages, we discourage putting more than one expression on a line.

Example: *Line Breaks*

The following lengthy expression is divided over six lines. We first break the expression at the assignment, and indent the code by two spaces. We next break the call to apply()

after the second argument, and align the third argument below the first. Finally, the closing curly brace aligns with the function definition and the closing parenthesis aligns with the apply() function call.

```
counts =
   apply(strategies, 1,
         function(strat) {
            sum(replicate(n, flipWhile(stratA, strat)) == 2)
         }
   )    ## apply function to strategies along the rows
```

Using line breaks is good practice for clarity in Python too, but you will need to use additional parentheses or brackets to signal that you are continuing across multiple lines.   ∎

5.3 Refactoring Code

As with writing, the editing process is an important part of coding (see Chapter 10). Our first code drafts can greatly improve when we abstract small pieces of code into reusable functions (and methods), parameterize code so that it is loosely coupled, remove redundancies, and break large tasks into smaller ones to make the code modular and less complex. Sometimes, it makes sense to discard our first version and rewrite from scratch in a simpler more straightforward design. The second time around usually goes more quickly because we have a better sense of how to organize the tasks.

Refactoring code is essentially an editing process, where we modify our code to consistently follow the style guidelines and general coding principles that we have adopted. This includes making the following changes to our code:

- replace hard-coded values with variables,

- rename variables and functions so they are more meaningful,

- format code with consistent indentation, spacing, and line breaks,

- break large functions (or methods) into smaller more focused tasks,

- create helper functions to carry out repetitive tasks, and

- update documentation to reflect the refactoring.

Example: *Refactoring Code for a Repetitive Plot Task*
In a simulation study, the simulation is run for various values of a parameter lambda. The results for a subset of these lambda values appear in Figure 5.1. Notice that some of the lambda values are placed on the plot near their associated curves. Locating the position to add the lambda values on the curve is in part trial and error and in part

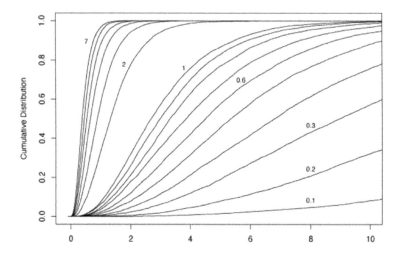

Figure 5.1 Plot Showing Subset of Lambda Values.

determined by a function. At first, we placed each label by repeating code seven times, once for each label.

To make the code easier to edit, we refactored it by gathering the arguments into vectors with informative names. Then we rewrote the code to loop over these values and place the labels on the plot.

```
indexes = c(1, 2, 3, 6, 10, 11, 16)
x_locs = c(8, 8, 8, 6, 4, 2, 0.75)
positions = c(3, 3, 3, 2, 2, 2, 2)

mapply(function(s, x, r, p)
  {
    Fn = ecdf(s)
    text(x=x, y=Fn(x), labels=r, pos=p, cex=0.8)
  },
  s=samples[indexes], x=x_locs,
  r=rates[indexes], p=positions)
```

Now when we want to add or remove a label, we can simply update the corresponding element in indexes, x_locs, and positions. Furthermore, if we decide to change other parameters, such as cex, we can do this once, not seven times. ∎

We refactor code to ensure its readability, and often as a side benefit we have code that is easier to understand, keep up to date, and extend.

5.4 Coding Principles

When we write about scientific findings, we follow some basic principles. We use precise language, make concise arguments, and organize our writing into sections and subsections to help the reader navigate our report and follow our arguments. Similarly, when we code, we follow conventions for layout and naming, and we abide basic coding principles so that our code is easy to understand and maintain. A key rule is the Don't Repeat Yourself (DRY) principle. DRY is at the root of important qualities of code design, such as abstraction and modularity, and it helps in writing code that follows the Keep it Simple (KIS) principle.

When we find ourselves copying and pasting chunks of code while only changing a few aspects each time, we should consider writing a function to do the job more parsimoniously. That is, we identify the common task, create a function to perform this task, and design the changing aspects as inputs to the function. This way, when we need to change the code, we update it in one place. Our code is more easily maintained and avoids having slightly different versions that become out of sync, contradictory, and error prone.

When a function (or any code) becomes too long, has too many purposes, or too many parameters, that is a signal to identify smaller tasks and create functions or modules with fewer inputs to carry out these tasks. Ideally, these smaller modules operate without the knowledge of another module's inner workings.

In cleaning and preparing raw data for analysis, we typically carry out several tasks, some of which we have not anticipated. We often backtrack and iterate parts of the data preparation as we come across new issues with the data in our analysis. As we clean and prepare the data, we organize the work into distinct tasks and wrap these tasks into functions that help pre-process the data. This way, when we discover a new task, it is easy to repeat part or all of the pre-processing. Below is an example.

Example: *Functions for Data Processing*
In an analysis of data that represent WiFi signals taken by stationary detectors in a building, the signals are measured repeatedly at various locations and orientations. The raw data are log files that do not have a simple format, and we need to read the rows as strings to separate them into fields. We then clean and transform the fields. This includes converting strings into numeric values and rounding values. Additionally, we discard extraneous rows of measurements, and we roll up multiple measurements taken at the same location and orientation into summary values. Below is pseudocode that identifies the various tasks and their inputs and outputs for cleaning and preparing the data.

```
raw_data = vector of n strings read from the data file

process_line(raw_data)
          extract 6 fields from each string that correspond to
          the x, y, z location, orientation, detector, signal
```

```
        Args:
                raw_data: n strings
        Returns:
                n by 6 matrix of strings

drop_extra_detectors(raw_matrix, which_detectors)
        drop rows in raw_matrix not from specified detectors

        Args:
                raw_matrix: n x 6 character matrix
                which_detectors: vector of detectors to be kept
        Returns:
                n by 6 matrix of strings

create_DF(raw_matrix)
        convert x, y, z, orientation, and signal to numeric
        and round orientation values

        Args:
                raw_matrix: n x 6 character matrix
        Returns:
                m x 7 data frame

round_orientation(orientation)
        round orientation values to nearest 0, 45, 90, ..., 315

        Args:
                orientation: vector of numeric orientations
        Returns:
                vector the same length as orientation

collapse_signals(signal_df, summary_func)
        group records by x, y, z, orientation, detector
        apply the summary function to the groups of records

        Args:
                signal_df: m x 7 data frame
                summary_func: fcn to summarize signals, e.g., mean
        Returns:
                k x 7 data frame
```

The separation of the work into these five tasks helps clarify the cleaning process, and wrapping the tasks into functions helps streamline and improve the process. ∎

Example: *Function to Find File Paths*
To build a data frame for a particular analysis, we need to read more than 9,000 files organized from five directories. Ideally, we would write code for this task that is independent of the location of the files on our computer. For example, we can place the files in a package where we know the path to the package, and programmatically construct the file names to read the data. For example, if the data are in the package called RSpamData, then we can find the path to this package with

```
spamPath = system.file(package="RSpamData")
```

Anyone who accesses the data from this package will have the identical organization of files.

In this example, the files are stored in five directories within the `messages` directory in the package. We can create a variable with these directory names (dirNames), and use list.files() to build the full path names for the files with

```
fullDirNames = paste(spamPath, "messages", dirNames,
                    sep=.Platform$file.sep)
## paste vectors together, separated by appropriate file separator

fileNames = lapply(fullDirNames, list.files, full.names=TRUE)
## apply list.files function to each element in fullDirNames
```

The first file name on our computer is

```
[1] "/Users/nolan/RPackages/RSpamData/messages/easy_ham/
       00001.7c53336b37003a9286aba55d2945844c"
```

Notice that we did not hard-code the path name for this file. Instead, we created it with the use of general function calls and knowledge of a common file organization. Another general approach is to use the local path from the current working directory to the data files. This approach can be useful when we distribute data and code together. ∎

5.5 Computational Reproducibility

Readers may want to see the implementation details of our code after reading the overview in our report. These details are not reflected in pseudocode and are not often the focus of reports and articles. Making our code available allows people to check our work and offer suggestions for how we can improve our analyses. When we make our code and data available, we are aiming for reproducibility. That is, we want others to run our code and get the same results (see Section 3.4 for more details).

When we work on a project, we often take a non-linear path, jumping around between ideas, and cycling through code multiple times. Organizing our code for sharing requires us to synthesize this process and identify the minimum code needed to reproduce our analysis and the figures in the report. Details such as how we cleaned the data (e.g., dealt with outliers and missing data) are often skipped in a report because it is not focal, but these choices can impact the outcome and need to be captured in the code and shared. Higher-level details such as the set-up of our computing environment are also important to share. We also provide our computer's model and operating system, the version of the software used, and, if needed, any additional libraries and their versions, and the random number generator and its seed (initial state).

A reader does not necessarily need to see everything we tried, nor the order in which we tried them. The same principles for organizing our findings described in Chapter 6 apply to organizing our computational process. That is, we determine what details are important for our report and ordered them in a compelling narrative. We should include code that produces the details mentioned and any other code needed to reproduce the results presented. Ideally, we would sequence code to match our report's storyline, but that is not always possible. Secondary analyses that do not contribute to the results discussed in our report are not necessary to include in the code we share.

We always aim to write efficient code, but often fall short. Needing to share our code can give us the push to improve and clarify code that is merely functional to code that is efficient and we are proud to share. However, the need for transparency trumps beauty, meaning if the option is between sharing "ugly" code and not sharing any code, we cannot let the fear of others seeing our less than perfect code prevent us from making our work available.

5.5.1 Version-Control Tools

We have had the experience where we saved multiple copies of the same script and incremented a draft number or date in the filename to keep track of our progress. Unfortunately, this approach makes it easy to lose track of things and start working on the wrong version. Version-control systems can help; they automatically track changes that we make to our files so that we can easily go back to previous versions without the onerous, confusing, error-prone clutter of storing each individual version "by hand". This approach is also useful if we are trying out something new and accidentally break the code. With version control we can always revert to an earlier version. Versioning systems can also be helpful for writing articles because we can return to previous drafts.

Version-control is especially useful in collaborations. Collaborators can make changes locally and then make their changes available to others by syncing their work. This ensures that team members do not override one another's work. Version control tools also make it easy to track changes between different versions so that you do not have to read through all of the code to see what a collaborator has changed.

5.5.2 Notebook Tools

Tools such as RMarkdown and Jupyter notebooks enable us to include code and results from that code, interspersed with text to help narrate the computational process and ensure that there are no copy/paste errors when moving results into a report. These dynamic tools can make production of a report seamless because the report is written with the same interface used to write code. When we change the code, figures, tables, and other outputs in the report are updated automatically. This process ensures that our report reflects our work and minimizes errors that come from mistakes in copying and pasting new results into a report draft.

5.6 Notes

Coding resources abound on the Web. The resources provided here are by no means exhaustive but should be a good starting point to delve more into pseudocode, coding principles, version control, and notebook tools.

The bootstrap pseudocode (Section 5.1) was adapted from Wasserman [18]. More information about pseudocode can be found in Viking Code School resources [17].

Our style-guide recommendations for writing code is adapted from "Google's R Style Guide" [5], "Google's Python Style Guide" [4], and from Pike [11] (the focus in this chapter has been on the R [13] and Python [16] languages). The origin of the DRY principle is from Hunt and Thomas [7]. The KIS (or KISS) principle originated in the US Navy (see Dalzell [3]).

There are many tools available to help keep track of your thought processes and make your work reproducible. These tools include ways to track different versions of code and drafts and ways to integrate text, code, and code output into one document. GitHub is a popular cloud-based service for version control. It is based on the version-control system Git, but others systems exist such as Concurrent Versions System (CVS), Subversion (SVN), Mercurial, Bazaar, and Monotone. Comparisons of version-control systems are made by Stansberry [15]. The limitations of Git are described by Williston [19]. Jenny Bryan and Jim Hester provide ample resources for using Git with R [2].

The Sweave tool [6] allows R code to be displayed and run where the output would be included in a compiled LaTeX document. The knitr R package [21] was developed to mimic Sweave but streamlines the process for those who do not know LaTeX [9]. RMarkdown [1] simplifies the syntax further, allowing users to make one document that contains different "chunks" whether the text is written in the simpler Markdown syntax. Jupyter notebooks similarly allow for a report written in Markdown to be created by combining these various elements for use with code in several languages, including Python and R. For more information about notebook tools in R consult Leisch [10], Xie [22], and Xie and Allaire [23]. More information about Jupyter notebooks can be found at the project's website [12]. Katins [8] provides knitr and RMarkdown capabilities for Python. For a comparison of R and Python notebooks, Woolf has a blog post [20].

For more practice, consider working on Portfolio Piece B.

5.7 References

[1] J. Allaire, Y. Xie, J. McPherson, J. Luraschi, K. Ushey, A. Atkins, H. Wickham, J. Cheng, W. Chang, and R. Iannone. *rmarkdown: Dynamic Documents for R*. London, 2020. R package version 2.2.

[2] J. Bryan and J. Hester. Happy git and GitHub for the useR. http://happygitwithr.com/.

[3] T. Dalzell. *The Routledge Dictionary of Modern American Slang and Unconventional English*. Routledge, 2009.

[4] Google. Google's Python style guide. http://google.github.io/styleguide/pyguide.html.

[5] Google. Google's R style guide. http://web.stanford.edu/class/cs109l/unrestricted/resources/google-style.html.

[6] P. Grosjean. *SciViews-R: A GUI API for R*. University of Mons, Belgium, 2014.

[7] A. Hunt and D. Thomas. *The Pragmatic Programmer: From Journeyman to Master*. Addison-Wesley, 1999.

[8] J. Katins. knitpy. https://github.com/jankatins/knitpy.

[9] L. Lamport. LATEX: a document preparation system, 1986.

[10] F. Leisch. Sweave: Dynamic generation of statistical reports using literate data analysis. *Compstat*, pages 575–580, 2002.

[11] R. Pike. *The Practice of Programming*. Addison-Wesley Professional, Boston, MA, 1999.

[12] Project Jupyter. jupyter. https://jupyter.org/, January 2020.

[13] R Core Team. *R: A Language and Environment for Statistical Computing*. R Foundation for Statistical Computing, Vienna, Austria, 2013.

[14] K. Schoeler. Data visualization & choosing a colorbrewer palette from an interactive app. https://stat198-spring18.github.io/blog/2018/04/24/choosing-a-colorBrewer-Palette-Interactively, April 2018.

[15] G. Stansberry. 6 version control systems reviewed. https://www.smashingmagazine.com/2008/09/the-top-7-open-source-version-control-systems/, September 2008.

[16] G. Van Rossum and F. L. Drake. *Python 3 Reference Manual*. CreateSpace, Scotts Valley, CA, 2009.

[17] Viking Code School Prep. What is pseudo-coding? http://www.vikingcodeschool.com/software-engineering-basics/what-is-pseudo-coding.

[18] L. A. Wasserman. *All of Nonparametric Statistics*, chapter 3. Springer, New York, 2006.

[19] J. Williston. When Git isn't good enough: Version control for enterprises. https://www.infoworld.com/article/2955650/development-tools/git-isnt-good-enough-version-control-enterprises.html, August 2015.

[20] M. Woolf. Advantages of using R notebooks for data analysis instead of Jupyter notebooks. http://minimaxir.com/2017/06/r-notebooks/, June 2017.

[21] Y. Xie. *knitr: A General-Purpose Package for Dynamic Report Generation in R*. R package version 1.28, 2020.

[22] Y. Xie. *Dynamic Documents with R and knitr*. Chapman & Hall/CRC, London, 2014.

[23] Y. Xie and J. J. Allaire. *R Markdown: The Definitive Guide*. Chapman & Hall/CRC, 2018.

5.8 Activities

5.8.1 Writing Pseudocode for Simple Statistical Tasks

Write pseudocode for one of the following tasks. Make sure your pseudocode identifies the inputs needed and uses active verbs to describe the computational task. Your pseudocode should be written in plain English and use indentation for nested procedures.

- Obtain the uncertainty of the mean of a collection of numbers via the bootstrap.
- Divide a dataset at random into a train and test set.
- Use k-fold cross-validation to estimate mean square prediction error of a model.
- Classify a new observation using k-nearest neighbors.
- Use stepwise selection to choose covariates for a model.

If possible, find a friend to read through your pseudocode and comment on whether or not they think they could implement the procedure in the programming language of their choice.

5.8.2 Mapping a Computational Task with Pseudocode

Read through the R or Python code shown below and determine what it does. Write a few lines of pseudocode that capture the basic computational task performed by the code. Edit your pseudocode to remove unnecessary details and clarify the main computational procedure.

R Version

```
flipWhile = function(strat1, strat2=c("H", "H", "T"))
{
  winner = NA
  flips = sample(c("H", "T"), size=3, replace=TRUE)
  while (is.na(winner)) {
    if (all(flips == strat1)) {
      winner = 1
    } else if (all(flips == strat2)) {
      winner = 2
    }
    flips = c(flips[2:3], sample(c("H", "T"), size=1))
  }
  return(winner)
}
```

Python Version

```python
import numpy as np
def flip_while(strat1, strat2 = np.array(["H","H","T"])):
    winner = float("NaN")
    flips = np.random.choice(("H","T"),
            size=3, replace=True)
    while (np.isnan(winner)):
        if sum(strat1 == flips) == 3:
            winner = 1
        elif sum(strat2 == flips) == 3:
            winner = 2
        flips = np.array([flips[0],flips[1],
                np.random.choice(("H","T"),
                size=1, replace=True)[0]])
    return winner
```

5.8.3 Verifying Reproducibility

When we check our code for reproducibility, it can be easy to miss aspects of the code that may not work on someone else's computer. The best way to confirm that our code is reproducible is to pass along our data and code to a peer and have them try it out.

Take some code and data that you have used for a project or assignment, give them to a peer, and see if they can run your code without any errors or unexpected warnings and get the same results as you did.

Watch out for the following potential pitfalls:

- differing version number of the software or libraries,

- path names to files that are specific to your computer,

- failure to install and/or load needed libraries,

- hidden dependencies created by your running code chunks out of order, and

- settings and preferences built into your software session.

5.8.4 Practicing Code Review

In addition to identifying violations of the style guidelines with simple examples such as those in Section 5.8.6, it can be helpful to review real-world examples of code. Exchange code with a peer, and take turns reviewing one another's code and offering suggestions for changes. Follow these steps to carry out your code review.

1. Read through the code and ascertain what you think it is supposed to do.

2. Check the code adheres to the style guidelines of Section 5.2 or a similar set of guidelines. Is there a blank space after every comma? Is the code properly indented? Are there blank spaces before and after operators , e.g., +, *, etc.? Circle any problem areas on the code itself.

3. Check a function's signature against the code in the body of the function. Are there any global variables that are being used? Identify them, if any, by writing "Glb" next to them above.

4. Presumably there are no syntax errors, but give the code a look through for that.

5. Look for good coding principles, such as the DRY principle (see Section 5.4). Are there areas where the code can be simplified by employing better coding practices. For code written in R, check that calculations are vectorized. Make your suggestions for changes to the right of the problematic code.

6. Do you see code that might not execute as expected? You might have uncovered these already in previous steps, but if not, make your suggestions here.

7. Review the documentation. Now that you have a good sense of what the code does, confirm that the documentation matches the code. Is every function documented as described in Section 5.2.2? Identify where in the code it would help to have additional or clearer comments. Also note where the comments are too specific and should be more general.

5.8.5 Write a Blog Post: Vignette

To practice both writing for a broader audience and communicating through code write a blog post that acts as a vignette at the same time. Think of a library or package that you use frequently or one that you want to learn more about. If the library or package lacks strong documentation, your blog post can help the open source community. A vignette should explain the main functionality of the library or package and show how it works through examples. For example, a student of ours, Schoeler, talks about choosing a color palette interactively in a blog post [14].

5.8.6 Dirty Dozen Code Recommendations

These are tongue-in-cheek recommendations which yield code that deviates from good programming practices. We provide snippets of R code that follow these "dirty dozen" guidelines and ask you to fix the code. We provide explanations of R-specific functions for Python users.

1. Name objects with similar, misleading, or uninformative names.

```
z1 = rnorm(10) ## draws ten random numbers from standard normal
z2 = rnorm(10)
z3 = rnorm(10)
```

What is a better approach to naming these three objects? Create one object, called z to contain all three vectors. What is the most suitable data structure to use? What if the three calls to rnorm() each generated a different number of random values?

2. Ignore the type of a vector.

```
a =  letters[1:8]
a[2] = TRUE
```

Explain why the above code might introduce unexpected behavior later on in your workflow.

3. Make multiple copies of a large object.

```
big1 = bigDF[ , 1]
big2 = bigDF[ , 2]
big3 = bigDF[ , 3]
newDF = data.frame(big1,  diff=big2 - big3)
```

Rewrite the code so that there are no multiple copies of the data.

4. Initialize a vector, and do it one element at a time.

```
x = vector("numeric", n)
for (i in 1:n) {
   x[i] = 100
}
```

Create the vector x without a loop (or apply function).

5. Subset illogically.

```
for (i in 1:n) {
   if (x[i] < -1) {
      y[i] = 0
   } else {
      if (x[i] > 1) {
         y[i] = 1
      } else y[i] = 3*x^2
   }
}
```

Rewrite this code, eliminating need for loops and if statements.

6. Rely on globally set variables inside a function.

```
myFunc = function(x) {
  x = rnorm(100)
  return(x + K)
}
```

Change the signature of the function so that it is clear which external variables are needed.

7. Apply a vectorized function to a vector one element at a time.

```
for (i in 1:n){
  y[i] = floor(x[i]) + 2 ## floor rounds number down
}
```

Rewrite this code without using a loop or apply function.

8. Put an operation that needs to be done once inside a loop.

```
for (i in 2:length(y)){
  k = 2
  y[i] = floor(sqrt(y[i-1])) + k
}
```

Rewrite the code to make it more efficient.

9. Run code several times over, each time tweaking it slightly.

```
# adds text and points to a plot
text(0.289, -1.225, "+500,000", cex=1.3)
points(0.239, -1.225, pch = 19, col = "red")
text(0.289, -1.245, "+100,000", cex=1.3)
points(0.239, -1.245, pch = 19, col = "red")
text(0.289, -1.260, "+50,000", cex=1.3)
points(0.239, -1.260, pch = 19, col = "red")
text(0.289, -1.270, "+10,000", cex=1.3)
points(0.239, -1.270, pch = 19, col = "red")
```

Write a function and include as arguments to the function the arguments of text() and points(), including those that do not change from one call to the next. Then write an expression to call your function. Recall text() and points() are both vectorized.

10. Keep every possible intermediate result, just in case.

```
z1 = myFunc(x)
z2 = z1 * 2
z3 = log(z2)
z4 = z3 + 10
```

Simplify these computations into one expression.

11. Leave objects in your workspace rather than saving them to an rda file.

If you are not familiar with different ways to store objects in your preferred programming language, do some research about options. Note any space or memory limitations.

12. Document your code incompletely. Keep similar-looking functions around with nearly the same names, some of which work and some of which don't.

Write two versions of a simple function, one that works and another that doesn't work as expected. Document these functions poorly. Give them to a peer and see if they can pick the correct one, point out the flaw(s) in the bad one, and update the documentation to the good one.

Part III
Composing the Story

Writing a first draft can be daunting. This part of the book provides strategies for getting started, templates for organizing material, and advice on what to write in different parts of a science article.

In Chapter 6, we demonstrate one approach to begin writing. Inspired by graphic organizers for laying out a video, we show how to use statistical graphs and tables to construct a storyboard. We start in the middle, with the results of our analysis in hand, and sketch out the arc of the story in our findings.

This organizational strategy helps us distill the results of a data analysis and identify and prioritize the main arguments in our work. We can then write high-level descriptions of our main points in the introduction. Storyboarding also helps us recognize the ideas on which the story relies. These main ideas form the core of the report's background section. In a discussion section, we write about important assumptions and shortcomings that surfaced during the research process. Through storyboarding we may even find a hole in our story that tells us we need to revisit the analysis or write about this hole as a topic for further investigation.

With a storyboard we can more efficiently outline a report and note connections between topics. In Chapter 7, we provide advice for writing the first draft starting from the storyboard. We begin in the middle and write the results section. From there, we proceed to the end and write the discussion and conclusions (these include points that surfaced in the storyboarding process). Then we circle back to write the beginning. This nonlinear drafting process is not the only approach. However, we have found it especially useful for beginning writers.

Some of the activities in Chapters 6 and 7 contain suggestions for how to draft press releases and blog posts, and they provide practice in storyboarding and writing for broader audiences.

6 Organizing the Story

Before we start writing a full draft, it can be helpful to organize both the structure and the story. In this chapter, we discuss how to build a narrative, select relevant details, and order them in a compelling way.

6.1 Creating a Storyboard

In the film industry, a storyboard is a series of sketches that depict the important changes of scene and action. These are arranged in a sequence to visually layout the progression of the story. Although storyboards are traditionally used as a means to brainstorm and plan out movies, TV shows, and other visual arts, the storyboard is fundamentally a tool to organize one's thoughts and streamline them into a story that is accessible and compelling. Data scientists too can benefit by mapping out their results into a storyboard before beginning to write.

We often start a data analysis with summary statistics and simple plots to understand the data. We then dig deeper to build and test models. Throughout the exploratory and formal data analysis, we make more plots and do more analyses than we want or need to display in a paper. Choosing what is most relevant to our argument is a valuable skill, and being economical is key. Determining the order to present our findings can be challenging because this order is often different from the sequence in which we performed the analysis. Creating a storyboard helps bridge the gap between our knowledge and the knowledge of our reader. Remember that you have spent much more time with the material than your reader so it can be difficult to pare down your findings and insights to those most relevant to your story. On the other hand, it can be easy to omit details that you are overly familiar with or think obvious.

We have organized storyboarding into six basic steps.

Communicating with Data: The Art of Writing for Data Science. Deborah Nolan and Sara Stoudt, Oxford University Press (2021). © Deborah Nolan and Sara Stoudt.
DOI: 10.1093/oso/9780198862741.003.0006

1. *Collect tables and plots.* In your work, you have most likely made many preliminary tables and plots to help you understand your data and models. Gather these tables and plots all in one place and in a format where you can easily move them around.

2. *Group related findings.* Group the tables and plots that contain similar messages, and summarize each group by one or two sentences or bullet points. Some groups will naturally form, but also consider regrouping and rearranging the plots and materials to uncover themes that are not immediately obvious. You may want to duplicate some plots and tables to put them into more than one group.

3. *Make an argument (find the story).* Consider the sets of tables and plots that you constructed and ask what they are telling you. Look at connections between groups and see if a story emerges. Sequence the groups so that they tell your story.

   The groups and the tables and figures within them may serve different purposes. One set of plots might build off of one another. Identify the train of thought: what is the starting point, what is the end, and what intermediate steps are needed to connect the two?

   Alternatively, plots in a group might supplement one another or supplement another set of plots. Determine how these plots connect to the main story. Do they branch off from an initial line of inquiry? Do they support a core argument?

4. *Choose the tables and plots needed to tell your story.* Be sparing with your plots and tables; choose the relevant details that explain the story while removing redundancies and unnecessary lines of inquiry. However, make sure that your core argument can be seen through your plots and tables.

   - If your plots present similar ideas, then decide which single plot or table suffices to explain or clarify the point. Remove the rest; they are redundant.
   - When you have material that builds off of one another, decide which intermediate steps, if any, are necessary for a reader to understand how to get from the start to the end of your story.
   - Identify holes in your argument that need a plot to support them, and make and add them to your storyboard.
   - In the case where materials supplement each other, decide if the supplementary information is necessary for your main storyline or if these details can go in an appendix or be summarized briefly in writing, and not shown with plots.

5. *Sequence the chosen tables and plots.* This requires a sense of fluidity between ideas. Examine all of the groups together and consider how each group relates to the others. Is there a temporal component between them? Does one group motivate another? Decide which details you need to see first to understand the

details that follow (e.g., would A make sense without first seeing B?). Think about how the supplemental information that survived the parsimony step ties back to the next major detail in the storyline. Can it be placed after the main storyline as a discussion point or before to set the stage?

Remember that the order that best expresses your ideas in a story does not necessarily match the order that best expresses your research process. A fluid order most likely does not match the chronology of your analysis.

6. *Add captions and transitions.* Write a brief caption for each plot that conveys the message you want the reader to glean. An important consideration in this step is whether, for the material being summarized, there is a mismatch in experience between you and your eventual audience. It can be easy to forget that the reader only knows what is being presented, and details were necessarily left out as part of the parsimony goal above. Is there any information that you could add between plots and tables that would smooth the transitions between them or provide context to ease interpretation?

Think of storyboarding as a visual outline that informs a formal written outline. We cannot write this formal outline without having an idea of what we are trying to say. The main goal of a visual outline is to identify the narrative. What problem exists? What did we do to solve that problem? Why does it matter that we have solved this problem? After we have identified the story and experimented with the ordering of details by rearranging panels of plots and text summaries, we can build a formal outline to tell the story we have identified. An intermediate step is to take notes on your storyboard.

6.2 Taking Notes on the Storyboard

The storyboarding process helps you separate the journey you have made to arrive at your findings from the findings themselves and prepare to write. Note-taking is another step in this direction. Before jumping into writing a first draft, take the opportunity to look over your storyboard and make notes. Reflect on the storyboard and jot down information needed to support your argument. These notes should consider the following topics.

1. *Foundations.* Make a list of the concepts that are most important to your findings. For each, determine whether it is: a concept that needs to be explained before presenting your findings; an idea that your work extends or that your work fundamentally relies on; or a motivating or competing notion that needs to be mentioned.

2. *Assumptions.* Augment your notes with any assumptions you made in your analysis. Annotate them with a brief indication of why they are reasonable, how they were confirmed, and what went unchecked.

3. *Eliminations.* Examine the plots and summaries that were dropped from your storyboard. Why were they dropped? Do they warrant mention in the report? If so, then add them to your notes. Write down whether the point should be included before, concurrent with, or after the findings. Also, if appropriate, mention where supplementary material on the topic can be found, e.g., an online resource or an appendix that you plan to write.

4. *References.* Undoubtedly, you have read many articles that relate to your work. Look over your findings and identify a few of those articles that are most relevant to your story. These can include an article that your work extends, one that justifies the method you have used, or one that supplements your findings.

Taking notes on your storyboard will help you settle on an outline for your narrative and solidify your argument. These notes (together with your storyboard) form the basis of the introduction, background, motivation, and discussion sections of your report. They are not meant to be exhaustive. Instead, their purpose is to identify the supporting material needed to tell how you arrived at the findings in your story. With the storyboard fresh in your mind, the core ideas, terms, and references are readily apparent. Noting them down now will save time when preparing your first draft.

6.3 Iterating

Storyboarding is part of the cycle of discovery. As we annotate plots, write brief transitions between them, and take notes on the storyboard, new ideas often occur to us, and we find gaps in our argument. With the help of the storyboard, our core arguments surface from our plots and tables, but as the essential pieces come together, we often discover that the justification of a point is missing or that a new line of inquiry appears. When this happens, we find ourselves updating our analysis, adding to our storyboard, and taking more notes.

Neither a storyboard nor its initial summary are fixed, and we don't always respond in the same way when we identify something missing. There are three basic approaches that we take to address newly found holes in our argument.

1. *Acknowledgement.* A hole does not always need to be filled, but at a minimum it needs to be acknowledged. In the report, this acknowledgement could be an identified assumption, a point of discussion, or a topic of future work. We update our notes to this effect.

2. *Patchwork.* The hole may have a quick fix, e.g., update a plot, run an additional simulation, skim articles on a related subject, or run-through a variant of the analysis. This new work may be included in the storyboard, or it may simply result in an additional note that summarizes the new information.

3. *Rework.* When we think the new issue needs to be more fully investigated, then we take a more thorough approach in revisiting the analysis. For example, we

may need to carry out a parallel avenue of investigation or check out a special case. If we believe it will strengthen our argument to flesh out this new idea, then it's best to tackle it at this stage.

Iteration happens throughout the writing process. After we have a storyboard and notes, we prepare a formal outline. Even this part of the process is iterative and can send us back to the data and our analysis. As we write up the essential pieces of our report, we may again find a hole and need to iterate. Our continued willingness to iterate will further strengthen our argument.

It's common for holes in your argument to crop up and lead you to retrace your steps. That's why we storyboard. However, when we are thorough in the early stages, the iterations in report-writing are likely to uncover issues of the acknowledgement and patchwork types that won't require major rework and rewriting.

6.4 Creating a Storyboard for Drug-Related ED Visits

We demonstrate the storyboarding process with an exploratory analysis of data from the Drug Abuse and Warning Network (DAWN). Each observation in this dataset represents a drug-related visit to the emergency department (ED) of a hospital in the USA. Other research has investigated the drugs accidentally ingested by children and those drugs responsible for over-medication among the elderly. We have chosen to focus our analysis on these two age groups to learn more about their drug-related ED visits.

We walk through the steps of creating a storyboard (Section 6.1) from a collection of crude tables and plots that represent the kind of output that we typically work with in the exploratory phase of an analysis. The collection includes tables of percentages and rates, bar plots of univariate statistics on the general characteristics of the visits, and bivariate plots that reveal relationships between age and other factors.

Age	≤ 5	6–11	12–17	18–20	21–24	25–29
Percent	5.8	1.4	5.6	6.6	8.1	9.4

Age	30–34	35–44	45–54	55–64	65+
Percent	8.6	14.3	15.3	10.2	14.8

Table 6.1 Age Distribution of Drug-Related Admits to an Emergency Department.

1. Collect Tables and Plots Our brief, simple analysis begins with one table and several plots (Table 6.1 and Figures 6.1–6.6). The captions for these figures and table are brief descriptions of the information plotted. The process of ferreting out the message in plots and writing informative captions is the purpose of storyboarding. Over the next steps, we cull the plots, revise them, and possibly add more as we piece the story together.

Table 6.1 gives an overview of the distribution of drug-related emergency department visits by age. Figure 6.1 shows the distribution of the type of drug-related visits across age groups. The left plot in Figure 6.2 compares the age distribution by sex, and the collection of four plots on the right side of this figure examine the types of visits by sex for a subset of ages. Figure 6.3 shows where patients go after their visit, by age. The bar plot in Figure 6.4 examines the time of day of the visit, by age. The set of dot plots in Figure 6.5 shows where patients go when released from their ED visit by time of day across a subset of ages. The four bar plots in Figure 6.6 explore the types of cases per quarter of the year across four of the age groups.

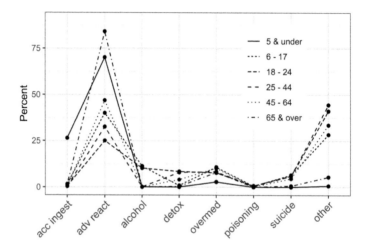

Figure 6.1 Line Plot of Case Type by Age. This line plot shows the proportion of cases within each age group; that is, values across types within an age group sum to one.

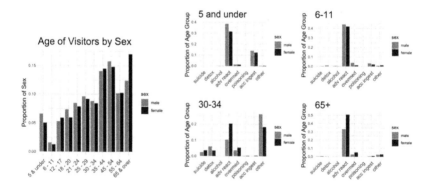

Figure 6.2 Bar Plot of Age by Sex. The plot on the left shows the breakdown of age for each sex; e.g., the sum of all bars for males is one. Each plot in the group of four on the right shows the proportion of case type by sex. Values within an age panel sum to one.

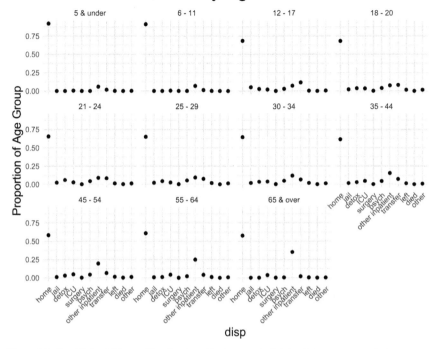

Figure 6.3 Dot Plot of Disposition by Age. Each panel of this plot corresponds to an age group. Within a panel, the proportions for each disposition, i.e., where the patient goes after the ED visit, sum to one.

2. Group Related Findings The patient disposition (where the patients go after their ED visit) appears most relevant to the elderly patients so we group together the set of dot plots in Figure 6.3 with those in Figure 6.5 that further break down disposition by time of day. We also group together Figure 6.4 and Figure 6.6 because they are most relevant to the youngest patients; that is, the time of day and seasonal differences are most striking for the five-and-under group.

We have grouped the plots into four sets. We have kept Figure 6.1 on the type of visits by age and Figure 6.2 on the age distribution by sex as separate sets. Figures 6.3 and 6.5 form a third set that pertains to the disposition of the elderly, and the last set consists of Figures 6.4 and 6.6 on time patterns in the visits of the youngest patients.

Other arrangements are possible. For example, we may want to duplicate the plots in Figure 6.5 and put them with Figures 6.4 and 6.6, as well as keep them in their original group. However, we don't bother to duplicate Figure 6.5 because these plots add little to the story.

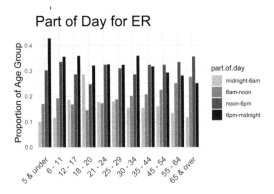

Figure 6.4 Bar Plot of Time of Visit by Age. This plot shows the proportion of each age group that arrives at the emergency department during different times of the day. The proportions within an age group sum to one.

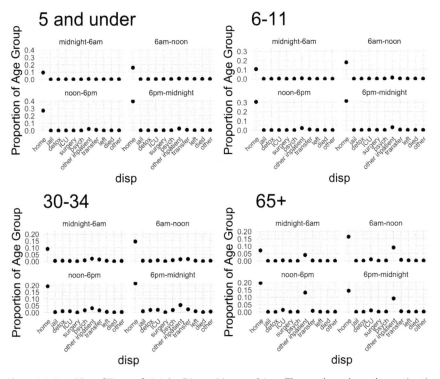

Figure 6.5 Dot Plot of Time of Visit by Disposition and Age. These plots show the patient's disposition by time of day. Each set of four plots represents a different age group: 5 and under, 6–11, 30–34, and 65 and older. Points across all four panels within an age group sum to one.

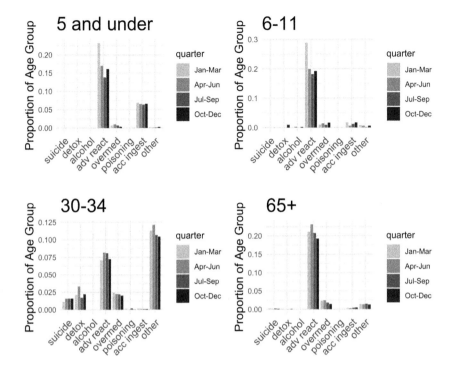

Figure 6.6 Bar Plot of Season by Type of Case and Age. This group of four plots show case type (x-axis) by the season for four age groups. The heights of all of the bars within a panel sum to one.

3. Make an Argument (Find the Story) We have found some important features of the ED-related visits for the youth (5 and under) and the elderly (65 and over). We chose to further focus on these two groups because our analysis may offer insights for informational campaigns that aim to reduce the frequency of ED visits for these two vulnerable groups. Specifically, we have found:

- These two groups make up a large share of the ED visits (more than 20%) and they share a large proportion of visits due to adverse reactions (at least 80% for each age group).

- For those five and under, the second-most common reason for a drug-related ED visit is accidental ingestion.

- For the elderly, the second-most prevalent reason is over-medication.

- Males outnumber females in the youth category, while the opposite is true for the elderly category.

- Youth visits occur most often at night (6 pm to midnight), and we see a larger percentage of youth visits during the winter months.

- After the ED visit, a noticeable proportion of elderly go to other in-patient care.

4. Choose the Tables and Plots Needed to Tell Your Story We can streamline Table 6.1 because it contains more age categories than needed, since our focus is only on the 5-and-under and the 65-and-older groups. A problem with this evidence is the lack of any comparison figures (see Section 3.3). For example, it could be helpful to compare the age distribution for ED visits to the age distribution of the US population. According to the 2010 US Census (Table 6.2), the proportion of youngest patients nearly matches the prevalence in the population, but the fraction of 6–11 year olds is less than one-fifth the Census figure. This comparison may offer insight. (Note that the Census data were provided for cruder categories so we collapsed categories in the DAWN data in order to make like comparisons between age groups.)

Source	5 & under	Age Group			
		6 to 11	12 to 17	18 to 24	25 to 34
DAWN %	5.8	1.4	5.6	14.7	18.0
2010 Census %	7.9	8.0	8.3	9.8	13.3

	35 to 44	45 to 54	55 to 64	65-plus
DAWN %	14.3	15.3	10.2	14.8
2010 Census %	13.3	14.6	11.8	13.0

Table 6.2 Age Distribution. The percentages of drug-related admits to an emergency department (first row) and 2010 US Census figures (second row) in each age group.

Next we consider the line plot in Figure 6.1 on the type of visit. This plot motivates the study and displays key features for both age groups of interest.

On the other hand, the plots in Figure 6.2 compare the age distribution (on the left) and the type of visit (on the right) by sex. Without a more in-depth analysis that compares our findings with the age-sex breakdown of the US population, we can't determine whether the differences in sex are noteworthy or simply reflective of population patterns. The advantage of comparing the sexes for the purpose of our analysis is not evident. We have identified an avenue of analysis that we don't wish to pursue, and we drop this plot from our storyboard.

The third group of plots (Figures 6.3 and 6.5) contain useful information about where elderly patients go after their visit to the ED. For now, we keep only the bottom three panels in Figure 6.3 because they contain the relevant information.

The fourth group of plots (Figures 6.4 and 6.6) concern the time of day and season when people visit the ED. The bar plot in Figure 6.4 shows the most striking pattern for those aged five and under so we prioritize it and drop the others.

In summary, we have selected Table 6.2, the line plot in Figure 6.1, the three panels of dot plots along the bottom of Figure 6.3, and the bar plot in Figure 6.4 to tell our story. Later, we will remake the plots to improve them.

5. Sequence the Chosen Tables and Plots The following sequence appears to be a natural progression of ideas.

- Start with the percentages from Table 6.2 to give context and a sense of scale.

- Then, describe the reason for the drug-related visits by age to motivate studying the youngest and oldest patients; use Figure 6.1.

- After we have established the vulnerability of the youngest, discuss when. Support from Figure 6.4.

- For the elderly, focus on the increase with age in the proportion of patients being admitted to the hospital. Base this on the bottom three panels in Figure 6.3.

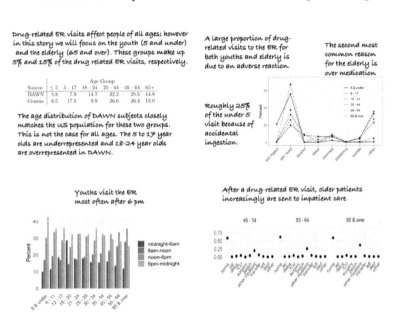

Figure 6.7 Example Storyboard. Here is an example of a storyboard coming together. Plots with a similar theme are gathered and summaries are added to them. From the original two tables and seven plots (with 38 panels in total), we winnowed down to one table and three plots to tell the story.

6. Add Captions and Transitions Figure 6.7 displays the flow of the tables and plots that we established in the previous step. That is, we put the figures and text together to form our storyboard. The captions and transitions shown in the figure reflect the thought process from the earlier steps in storyboarding. To practice creating your own, see the activities in Section 6.7.1.

6.4.1 Taking Notes on the DAWN Storyboard

With the fresh storyboard, we make notes of the supplemental information we expect to need for our first draft.

Our analysis relies on the the the soundness of the DAWN survey. Basic information about this survey is important for the reader to understand and believe our results. We want to identify records and how they were selected for the survey. Our analysis consists primarily of comparing percentages, and we will want to provide a measure of the accuracy of these percentage.

We eliminated the portions of the analysis that examined sex and season. We note these deletions and that we do not plan to include them as discussion points. This note serves as a record of our decision.

For references, we would list the source of the data and description of the DAWN survey methodology, and the source for Census figures. We would also list any additional references most relevant to the two age groups under study.

6.4.2 Iterate Over the Storyboard

		Age Group			
Source	5 & under	6–11	12–17	18–24	25–34
DAWN %	5.8	1.4	5.6	14.7	18.0
2010 Census %	7.9	8.0	8.3	9.8	13.3
Visits/100,000 people	1,201	288	1,107	2,455	2,217

	35 to 44	45 to 54	55 to 64	65-plus
DAWN %	14.3	15.3	10.2	14.8
2010 Census %	13.3	14.6	11.8	13.0
Visits/100,000 people	1,762	1,718	1,416	1,864

Table 6.3 Age Distribution for Drug-Related Emergency Department Visits. The top row displays the percentage of drug-related visits to the emergency department by age group, and second row gives 2010 US Census figures. The third row provides rates of the number of visits per 100,000 people in that age group. Most noticeably, the rate of visits for the youngest is four times the rate for 6–11 year olds.

We continue to review the storyboard and ask ourselves whether our argument holds together or needs additional work. For example, when we created the storyboard, we identified a hole in the lack of comparison figures for the age distribution and so added comparative statistics from the 2010 US Census (Table 6.2). As we iterate again, we consider whether the revised table adequately addresses the problem. What is the best way to illustrate the vulnerability of the youngest and oldest populations? Rates that normalize the number of visits by the size of the population are likely to offer a more informative comparison. We have identified a new hole in the hole that we patched earlier. For our new patch we add one more row to the table to provide the number of visits per 100,000 people in each age group (see Table 6.3).

We revisit our notes and add an item about the definition of the rates. Also, when we computed these rates, we found that the DAWN age categories don't match exactly the Census categories. Since we used approximations to make the age categories line up, we add a note about this.

After further contemplation, we deem the plot of time of day unnecessary because it doesn't focus adequately on our age groups. Instead, we decide to include some of the statistics from that plot rather than the plot itself. We make a note of this decision, but for the time being, we leave the plot in the storyboard as a placeholder.

In these iterations we continue to push our argument forward. We trim away unnecessary information, fill holes, and carefully prepare a defensible and convincing argument. Note-taking helps us keep track of the decisions that we make and why.

6.5 Notes

The idea for storyboarding our research was inspired by an activity led by Sara ElShafie in the Data Science for the 21st Century NSF Training Program Science Communication Short Course at UC Berkeley taken by Sara Stoudt. The activities described in this chapter are also inspired by exercises in this short course.

Examples of traditional storyboards can be found at The Walt Disney Family Museum in San Francisco and in the Khan Academy's exploration of Pixar's process [9]. A researcher's perspective on storyboarding research can be found at Dunleavy's blog [5].

The DAWN data used throughout Section 6.4 can be found at [15], and a description of the DAWN survey methodology can also be found there. The Census figures are available at [8]. Additional background references for the notes on the storyboard (Section 6.4.1) might include the Substance Abuse and Mental Health Services Administration (SAMHSA) [13], Bush [3], and Moharty and Slattum [11].

For more practice, consider working on Portfolio Piece G.

6.6 References

[1] American Beverage Association. Beverage industry responds to DAWN report on energy drinks, November 2012.

[2] J. A. Banken. Drug abuse trends among youth in the United States. *Annals New York Academy of Sciences*, 1025:465–471, 2004.

[3] D. M. Bush. Emergency department visits attributed to overmedication that involved the insomnia medication zolpidem. *The Center for Behavioral Health Statistics and Quality Report*, 2014.

[4] J. D. Colliver and A. N. Kopstein. Trends in cocaine abuse reflected in emergency room episodes reported to DAWN. *Public Health Reports*, 106(1):59–68, 1991.

[5] P. Dunleavy. Storyboarding research. https://medium.com/advice-and-help-\in-authoring-a-phd-or-non-fiction/story-boarding-research-b430cebd5ccd, 2018.

[6] G. Graff and C. Birkenstein. *They Say / I Say: The Moves That Matter in Academic Writing.* W. W. Norton & Company, New York, 2009.

[7] A. J. Hall, J. E. Logan, R. L. Toblin, J. A. Kaplan, J. C. Kraner, D. Bixler, A. E. Crosby, and L. J. Paulozzi. Patters of abuse among unintentional pharmaceutical overdose fatalities. *Journal of the American Medical Association,* 300(22), 2008.

[8] L. M. Howden and J. A. Meyer. Age and sex composition: 2010. *2010 Census Briefs,* 2011.

[9] Khan Academy. Pixar in a box: the art of storytelling. https://www.khanacademy.org/partner-content/pixar/storytelling.

[10] M. E. Mattson, V. A. Albright, Y. Yoon, and C. L. Council. Emergency department visits involving misuse and abuse of the antipsychotic quetiapine: Results from the drug abuse warning network (DAWN). *Substance Abuse: Research and Treatment,* 9:39–46, 2015.

[11] M. Mohanty and P. Slattum. Alcohol, medications, and older adults. *Age in Action,* 26(3):1–5, 2011.

[12] Substance Abuse and Mental Health Services Administration (SAMHSA). The DAWN report: Update on emergency department visits involving energy drinks: A continuing public health concern. https://www.samhsa.gov/data/sites/default/files/DAWN126/DAWN126/sr126-energy-drinks-use.htm, January 2013.

[13] Substance Abuse and Mental Health Services Administration (SAMHSA), Office of Applied Studies. *Emergency Department Visits Involving Accidental Ingestion of Drugs by Children Aged 5 or Younger.* Technical report, National Criminal Justice Reference Service, 2010.

[14] J. D. Swisher and T. Hu. A review of the reliability and validity of the drug abuse warning network. *International Journal of the Addictions,* 19(1):57–77, 1984.

[15] United States Department of Health and Human Services. Substance Abuse and Mental Health Services Administration (SAMHSA). Center for Behavioral Health Statistics and Quality. Drug Abuse Warning Network (DAWN). https://doi.org/10.3886/ICPSR34565.v3, 2011.

[16] Women's Health Research Institute. Energy drinks sending thousands to emergency rooms. https://www.womenshealth.northwestern.edu/blog/energy-drinks-sending-thousands-emergency-rooms, 2011.

6.7 Activities

6.7.1 Make a Storyboard

Use the plots and tables provided at the end of Section 6.7.5 to create your own storyboard about drug-related visits to emergency departments of hospitals. Univariate statistics that cover general characteristics of the visits such as sex and race of the patient, metropolitan area of the hospital, time of day when the visit took place, and case type are provided as well as bivariate plots that display characteristics according to three major themes: visit types, marijuana, and adults and aging. Pick one of these themes for your storyline. Consider printing the plots and using index cards or sticky notes so that you can physically group and rearrange the plots and tables. Alternatively, you can use software to produce slides that you can rearrange on your computer.

6.7.2 Peer Review a Storyboard

Begin by creating a storyboard (Section 6.7.1) and have a peer also create a storyboard. This activity is most useful if you choose different themes from one another.

1. Exchange storyboards. Can you interpret each other's story?

2. Try to understand the major take-aways of your peer's story. As you look at each panel in the storyboard, write a sentence or two about what you think each panel is saying. Use those sentences to write a paragraph that summarizes the story you think the storyboard is trying to convey.

3. Exchange your summary paragraphs and see if you understood one another's story. Seeing where your partner went astray shows you the weaknesses in your storyboard. What was unclear, and what could have been added to make the point more clear?

4. Refine your storyboard according to the feedback you received. Perhaps a particular plot is distracting from your main message or an extra transition sentence would help give context to the connection between two plots. The storyline might not be as compelling as you originally thought. How can you adjust the story to make it more accessible or exciting?

6.7.3 Find a Blog Thread in Your Work

When blogging about your work, you need to identify a small piece that is accessible and engaging. The following steps help you brainstorm blog post ideas. After you settle on a compelling idea, write a blog post outline.

1. Revisit your storyboard. Remind yourself what narrative fits with your work. Is there a part of the story that you could expand on? What part do you think is especially under-appreciated? Write down two potential angles for a blog post.

2. Think outside of the box. Writing a blog version of a formal report is not the only way to write a blog about your work. Having this goal in mind may limit your creativity. Instead, dig into a specific detail in your work or discuss your experience working on the project. Your blog post may be engaging for a new audience without focusing on your formal results. Write down two more potential blog post ideas that focus less on your formal work.

3. Get feedback from a peer. Talk to a peer (or someone who represents your intended audience) about your work and see what they think is exciting and what they are curious about. This step is important! Discussing your ideas with someone else can help you move from a vague idea to a concrete story, choose an effective angle, and help focus your blog post on a narrower thread of interest.

After your discussion with a peer, choose the blog-post idea that seems most promising. Outline the blog post.

6.7.4 Create a Storyboard for a Discussion Blog

Creating a storyboard can help us find where our work fits in the literature or where our point of view fits into a broader discussion. In this activity, you create a sample storyboard for a blog post that discusses the implications of the DAWN data and use the *They Say/I Say* framework [6].

This activity requires creating pieces of the storyboard describing both your work and the work of others to help you gain an appreciation for different viewpoints that make up a discussion. Providing transitions between those two types of storyboard panels helps you assess your argument in the context of a broader one. We provide an example below to help guide you.

I Say: Your Work The storyboard panels for what you say should follow the storyboard style shown in Section 6.4. However, you might want to talk about your findings at a higher level. What are the major themes? What do they mean in a broader context?

Swisher and Hu [14] discuss the reliability and validity of the dataset itself, so for the purposes of this activity we can use this information to add context about our own analysis of the DAWN data.

The studies reviewed in this section have indicated that DAWN data are both reliable and valid. The major problem with DAWN data appears to be that they underestimate the extent of the problem.

Recent reviews have assured that recording errors are only 54% on substances and 2–3% on sociodemographic information. The ranking and relative distribution of drug use have been found to be generally accurate and consistent. Therefore, for the purpose of formulating drug program, enforcement, prevention, and treatment policies, the DAWN data provide quite accurate information.

They Say: Others' Work To determine what "they" say, we need to do some research to see how others are discussing the DAWN dataset.

A Google Scholar search of the DAWN data yields literature that focuses on cocaine and opioids. This work rarely breaks things down by age, focusing more on aggregated trends over time. When we glimpse some information summarized by age, it is clear that our age groups of interest (youth and elderly) do not have a large number of emergency-room visits due to cocaine usage (see Colliver and Kopstein [4]).

However, there are some papers that overlap our populations of interest. For example, while studying misuse and abuse of a particular antipsychotic, Mattson et al. [10] uncovered some information about adverse reactions to the same antipsychotic. They found that elderly women have more ED visits involving adverse reactions. Hall et al. [7] studied death by unintentional pharmaceutical overdose in West Virginia. They found that men's drug-overdose death rate was more than twice that of women and the age groups with highest death rates were between 25 and 54.

We should note that there are some discrepancies in how we define terms. As an example, Banken [2] considers youths to be high-schoolers, studying students in eighth, tenth, and twelfth grade.

Putting it Together Once we understand the overall message of both our work and others, we need to find a way to stitch the two sides into one discussion. Consider the following questions to identify the relationship between what you say and what others say. What do the two sets of storyboard panels agree and disagree on? Does the emphasis differ between the two? Do the two span the space of views or is there still a component that is missing?

6.7.5 Make a Storyboard for a Press Release

For this activity you are to make a storyboard for a press release about emergency department visits related to energy-drink consumption. The following steps guide you through this process.

1. **Read a technical report.** Begin by reading about the analysis of energy-drink consumption made by the Substance Abuse and Mental Health Services Administration [12].

2. **Outline a storyboard.** Select the key points that you want to make clear in your press release. Remember to keep the material engaging, accessible, and brief. Use the inverse pyramid to organize your storyboard (see Section 2.1.3) and be sure to address the Ws (see Section 2.1.1).

3. **Add quotes from experts.** Consult the response from the American Beverage Association on related topics for relevant quotes [1].

When you complete your storyboard, read an example of a press release from the Women's Health Research Institute about the report and compare and contrast your approaches [16]. Iterate on your storyboard if their press release exposed holes in your argument.

Materials for Storyboard Activities

To generate materials for the storyboard activities in Sections 6.7.1 and 6.7.2 we use data from the Drug Abuse and Warning Network (DAWN) and generate output that mimics what we may obtain during exploratory data analysis. Each row in this dataset records a drug-related visit to the emergency department of a hospital in the USA. We generate tables and bar plots to display univariate statistics that cover general characteristics of the visits as well as plots to display bivariate relationships that shed light on three major themes: visit type, marijuana, and adults and aging. Using the plots and tables presented below, a storyboard can be created to try to express the storyline of one of the three themes.

sex		race		daypart	
male	47.64	white	68.55	midnight-6am	15.88
female	52.36	black	18.16	6am-noon	21.05
		hispanic	11.22	noon-6pm	31.67
		other	2.07	6pm-midnight	31.39

alc		illicit		pharma		nonmedpharma	
no	85.71	no	75.28	no	20.42	no	75.43
yes	14.29	yes	24.72	yes	79.58	yes	24.57

Marijuana		Amphetamine		Meth		Antidepressant	
0	95.96	0	99.34	0	98.90	0	97.31
1	4.04	1	0.66	1	1.10	1	2.69

quarter		metro		metro	
		bostonMA	1.89	nycNY	2.46
janMar	25.45	chicagoIL	2.89	detroitMI	1.87
aprilJun	26.55	minneapolisMN	1.07	fortlauderdaleFL	0.65
julSep	25.01	dadecountyFL	0.63	houstonTX	1.85
octDec	22.99	denverCO	0.72	phoenixAZ	1.24
		oaklandCA	0.65	sfCA	0.72
		seattleWA	1.07	other	82.29

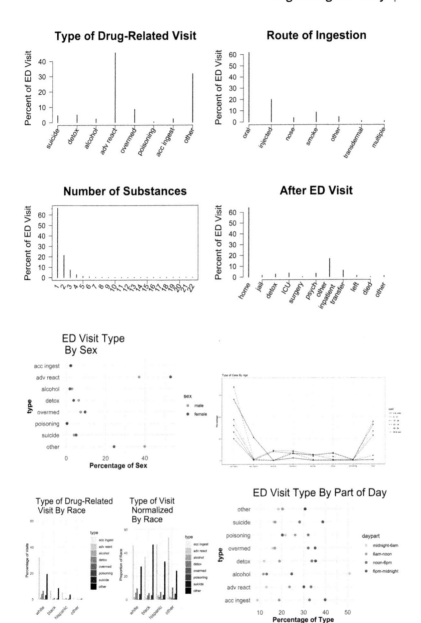

ED Visit Type By Quarter

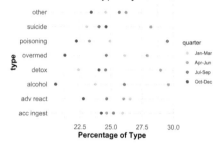

Marijuana-Related Visits By Metro Area

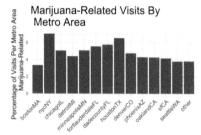

Marijuana-Related Visits By Metro Area and Sex

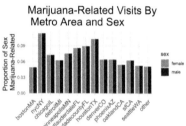

Marijuana-Related Visits by Age

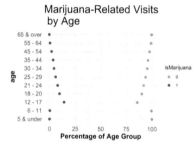

Marijuana-Related Visits By Age and Sex

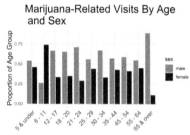

Marijuana-Related Visits by Race

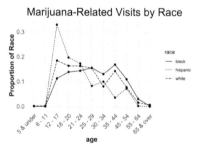

Marijuana-Related Visits

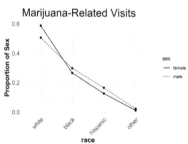

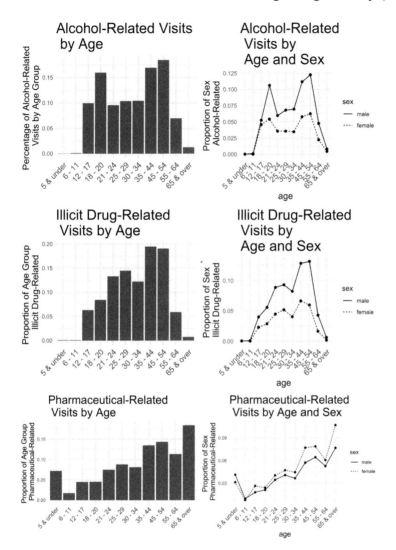

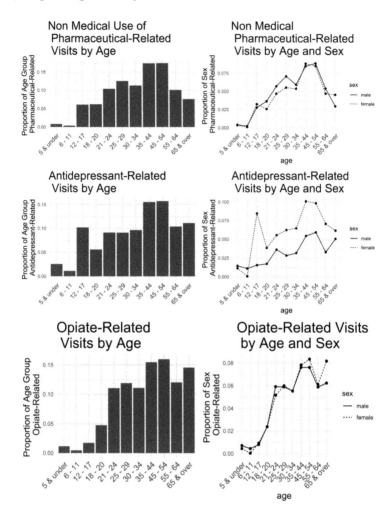

Race by Age

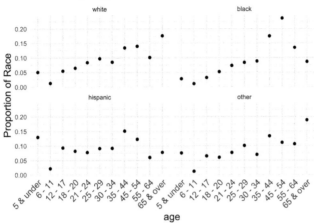

Location After Visit by Age

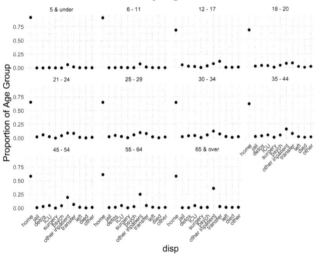

Location of Visit by Age

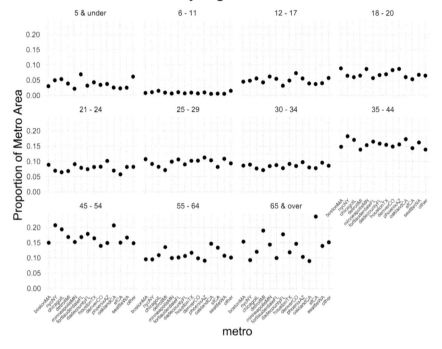

7 Writing the First Draft

Science articles about a data analysis have a typical structure that we introduce in this chapter. We discuss what should be included and emphasized in each part of the article and give advice for how to write the first draft.

An article is an extended argument, and as with any argument, there are two sides: yours and others. However, since you do not directly converse with the reader, you must foresee their potential objections, questions, and comments and address them ahead of time. As well as anticipating possible pitfalls, it is also important to anchor your claims to that of others. You can be extremely convincing in the moment, but if readers don't know what to compare your argument to, they may not remain convinced when presented with other work that contradicts yours. In this chapter, we provide examples of how to preface your work with what others have written on the topic, distinguish what you have found from what others are saying, and defend against skeptics.

As we describe the various pieces of a report, we consider three publishing venues, the *Morbidity and Mortality Weekly Report* (*MMWR*), *PLoS ONE*, and the *Journal of the American Statistical Association* (*JASA*). We use the storyboard we created in Section 6.4 to draft a brief report that follows *MMWR* guidelines. For the other two publications, we examine a sample article and consider how the authors have organized their article and made their argument.

7.1 Structure: What Should My Report Look Like?

In writing about a data analysis, the components of a report and the order in which they are put together tend to have a common structure across different types of publications. There may be more leeway in how to organize an article for certain publications, so before you start writing, look into the publication guidelines for the journals you want to send your work to and learn what structures these journals require or recommend. If you are writing a report for a course, then you will want to review the requirements

Communicating with Data: The Art of Writing for Data Science. Deborah Nolan and Sara Stoudt,
Oxford University Press (2021). © Deborah Nolan and Sara Stoudt.
DOI: 10.1093/oso/9780198862741.003.0007

given by the professor, or if you are writing your work for a project manager or supervisor, then discuss expectations with them and if possible ask for sample reports.

Example: MMWR *Organization*

The *MMWR* author guidelines provide detailed specifications for each of the seven types of articles they publish. We examine the information provided for writing a "full report". Full reports are limited to 1,400 words, and they must include a summary box that provides one to two sentence answers to the following questions. What is already known on this topic? What is added by this report? What are the implications for public-health practice?

The main body of the report begins with a one-paragraph introduction that specifies what the problem is, who worked on it, when and where, why the problem is important, and how it was addressed. The introduction is also expected to include the data sources, methods used, and a couple of key findings. Following the introductory paragraph, an additional paragraph can be provided that contains background information needed to understand the analysis and its implications.

A one or two paragraph description of the methods used to conduct the analysis follows the introduction. In addition to the statistical methods used for the analysis, this description includes, as appropriate, data sources and collection methods.

Next, the results section provides a concise description of the major results of the analysis. These results can include descriptive elements (i.e., time, place, person) and epidemiologic results, such as disease trends. The author guidelines suggest that any numeric values highlighted in the text should also be presented in a table or figure. Additionally, the report may only contain a combined total of three boxes, tables, and figures.

Finally, the discussion section states the conclusions of the report and conveys their public-health meaning. The discussion also must include a paragraph on the limitations of the analysis and conclude with recommendations for public-health officials. ∎

Example: *PLoS ONE Manuscript Requirements*

The Public Library of Science (PLoS) is an open-access publisher and *PLoS ONE* is its peer-reviewed journal that covers primary research in all science disciplines. The guidelines for contributors state that the manuscript should begin with a title page that contains the title, authors and affiliations, an abstract, and introduction. Then, the rest of the manuscript must contain sections about materials and methods, results, discussion, and an optional conclusion. The section titles and ordering is flexible. Figures and supporting information should not be included in the manuscript, but are submitted separately. However, figure captions are to be placed in the manuscript. ∎

Example: JASA *Applications and Case Studies*

Articles published in *JASA* fall into one of three main categories: statistical applications and case studies; theory and methods; and reviews. Aside from instructions on the abstract, no information is provided for how to organize an article. In such a situation, we examine a few published articles for guidance.

For example, "A Model-Based Approach to Climate Reconstruction Using Tree-Ring Data", by Schofield et al. was published in *JASA* in the Applications and Case Studies section of the journal. Their article has the following sections.

1. *Introduction.* Motivates the work, provides background material on other approaches, and ends with a brief description of the organization of the rest of the paper.

2. *Data.* Provides information about where and how the data were collected, a defense of choices made about which data was used in the modeling, and particular issues related to the data that must be addressed in the analysis.

3. *Standard Methodology.* Gives an overview of the current main approaches to the problem.

4. *Problems with Existing Approaches.* Points out the flaws in existing approaches and motivates the current work that aims to mitigate these problems.

5. *Model-Based Description.* Presents new models and describes how they are fitted and applied.

6. *Model Checking.* Explains how to evaluate the different models.

7. *Discussion.* Summarizes the merits of the new approach, discusses aspects of the analysis that could be expanded on, and concludes with a take-away message for practitioners in the paleoclimate community.

■

7.2 Audience: For Whom Am I Writing?

Knowing who reads our article, why they read it, and how helps determine the way we should write. Consider the knowledge your audience has about the subject. You want to tailor your terminology and the balance of background information to depth of analysis accordingly. A knowledgable audience may be bogged down by background information they already know, but an audience new to the topic may require some general set-up in order to understand your specific work.

The motivation of our audience is another consideration. If someone is reading to help understand a decision we are advocating for, we will emphasize our conclusion and the main details supporting it. If someone is reading to understand the state of the art and its limitations, we want to emphasize what makes our approach different and where there is still room for improvement.

How our audience reads our document is another factor. Most readers first skim an article in order to orient themselves, decide whether or not to read the article, and select the section(s) of greatest interest to them. That is, they may read the abstract and skim the introduction and conclusion, then page through to read the section titles, look at the figures and read the captions. Since readers are likely to jump around, it may be useful to frequently review what you have previously said and where you are heading to help orient them. No matter the goals of your reader, good section headings (and sub-headings) help orient them and make it easier to track the portions of your report that most interest them.

Before starting to write, make sure to acquaint yourself with the interests and knowledge of the readers you want to reach. Often, academic journals specify a scope, and this can help find a good match for your intended audience.

Example: Significance *Audience*

The magazine *Significance*, in which the article about donkeys in Kenya appeared (see Section 1.5), has a broad audience with varying levels of statistical expertise. The request for contributions explains that *Significance* is not a traditional research journal and contributions must be written in an accessible and engaging style. Technical terms and mathematics should be used sparingly, and when they are used, they must be explained clearly and for a nontechnical audience. ∎

Example: MMWR *Audience*

The audience for the *MMWR* Weekly includes the public-health community, private clinicians, researchers, teachers and students, and the news media. The submission instructions for authors states that *MMWR* reports should be concise and written in plain English with a minimal use of acronyms. The *MMWR* publishes seven types of reports, including "full reports" and "policy notes". The goal of a full report is to summarize an analysis that addresses an important public-health question and make recommendations. Timeliness of reports is important and requirements are placed on the age of the data. For example, a report that analyses surveillance data, such as the tobacco-use survey described in the examples in Section 3.1.1, must be submitted within three years of data collection. ∎

Example: JASA *Audience*

The Applications and Case Studies section of *JASA* has a wide audience within the statistics community. Since this section contains articles that analyze real data from a broad range of applied fields, we cannot assume the reader has any familiarity with the context of the data. We can assume that the audience has a statistical background, but we should provide information about the motivation of the analysis from the domain scientist's point of view to help the audience understand why the analysis is important. If the methods are fairly standard, we would not need to go into great detail for the *JASA* audience. However, if the methods are new, this audience will expect a thorough description of what makes the new approach different and better than the current state of the art. ∎

Example: PLoS ONE *Audience*

PLoS ONE publishes original research within any discipline in science and medicine. By including research from many subject areas, the journal aims to facilitate connections between researchers in different disciplines as well as across sub-disciplines. Since the audience expertise is quite broad, it is important to define key terms and methods specific to our field and to connect our approach to analogous work in other fields. ■

7.2.1 Secondary Audiences

Although we have a main audience in mind while writing a formal report, in most cases a report does not have just one audience. We recommend that you choose a primary audience to prioritize when writing your first draft, but when you begin the revising process (Chapter 10), it can be helpful to consider a secondary audience and write to also satisfy them as well. The secondary audience is often made up of readers outside of your particular field, and if you can keep this alternate audience in mind while writing, you will increase the accessibility of your work, which will ultimately help you broaden your impact.

7.3 The Middle

It may seem strange to start writing a draft in the middle, but having this part written makes it easier to synthesize and summarize our work at the beginning and end of the article. The middle contains the main content: what we did, what we found, and what our results mean. This often comes in the form of descriptions of the data, our methods, analysis, and findings, as well as accompanying figures and tables.

A common challenge for writing these middle pieces is to be thorough and specific while also being concise. Additionally, choosing the right balance of describing what we did versus what we found versus what our results mean in context also depends on the audience. For example, we may need a more detailed description and defense of our methods, if we are proposing a new or alternative method for analyzing some data or if we are aiming for a more technical audience. Other writing choices that depend on the final destination of our work and heavily impact how we write the middle include the length and pacing of the writing. Keep in mind that the length of your report is often driven by the length of the middle section. We tailor our writing to fit the space requirements of wherever we want our writing to be published.

7.3.1 Starting Point–The Storyboard

The storyboard we created from our statistical graphs and tables in Section 6.4 serves as the starting point for writing the middle of our paper. Recall we used the storyboard to sift through graphs and tables and organize them into a coherent story, and we also made notes of key assumptions, concepts, and points. Before we begin writing the middle of our report, we again revisit our storyboard (see Figure 7.1) and do the following:

Drug-related ER visits affect people of all ages; however in this story we will focus on the youth (5 and under) and the elderly (65 and over). These groups make up 5% and 15% of the drug related ER visits, respectively.

Source	≤ 5	5 - 17	18 - 24	25 - 44	45 - 64	65+
			Age Group			
DAWN	5.8	7.0	14.7	32.2	25.5	14.8
Census	6.5	17.5	9.9	26.6	26.4	13.0

The age distribution of DAWN subjects closely matches the US population for these two groups. This is not the case for all ages. The 5 to 17 year olds are underrepresented and 18-24 year olds are overrepresented in DAWN.

A large proportion of drug-related visits to the ER for both youths and elderly is due to an adverse reaction.

The second most common reason for the elderly is over medication

Roughly 25% of the under 5 visit because of accidental ingestion.

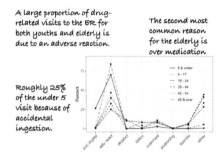

Youths visit the ER most often after 6 pm

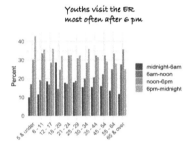

After a drug-related ER visit, older patients increasingly are sent to inpatient care.

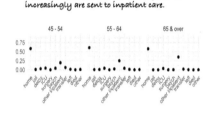

Figure 7.1 DAWN Storyboard from Section 6.4.

1. Look again for holes in our argument. Do we need to create another plot to bring out an important aspect of our analysis, or should we run another what-if analysis with our data to further justify our approach?

2. Remake any statistical graphs so that they best convey our argument. The graphs that we first made helped us formulate our narrative. Now that we have our storyline figured out, it is a good time to revisit these graphs to see if we can improve them to make our points clearer and more compelling.

Preparing an outline of our article from the storyboard and notes is a natural place to begin writing. This outline can serve as a guide to organizing the middle of our document and as a starting point for our introduction, background, discussion, and conclusions. We don't create an outline simply because we first need to describe what goes into each of the article sections.

Example: *DAWN Analysis Storyboard*
Can we gain insight into how to prevent drug-related emergency department (ED) visits among young children and the elderly? Using the Drug Abuse and Warning Network (DAWN) data we can understand patterns in, reasons for, and outcomes of drug-related emergency visits for these vulnerable groups.

In Section 6.4, we created a storyboard on this topic. After following the steps given in the chapter, we crafted a storyboard with the following points.

- The five-and-under and 65-and-over patients together make up 20% of the patients.

- The rate of emergency department visits among children aged five and under is four times the rate for children aged 6–11 years old.

- Nearly all visits from those aged five and under are due to either adverse reaction or accidental ingestion, and one in four of the cases of accidental ingestion is due to the ingestion of pain medication.

- The elderly visits are also primarily due to an adverse reaction, with over-medication a distant second.

- Those aged five and under visit emergency departments after 6 pm and nearly all return home after the visit.

- The elderly typically visit before 6 pm and often require follow-up with other in-patient care.

The plots that conveyed our story are shown in Figure 7.1.   ■

7.3.2 Data-Collection Methods

A discussion of the data and methods may seem fairly standard for a quantitative report, but the framing of these sections depends on what aspect of our work is considered novel. There are two basic ways our analysis can be novel:

- *Newly Collected Data, Traditional Statistical Methods.* The novelty here comes from the data, i.e., with these data we have the ability to answer new questions using standard methods of analysis. When the data provide the novelty, we must emphasize why this particular dataset is important. What questions can these data help answer that were previously unanswerable? What information was lacking in datasets similar to ours?

- *Known Data, Newly Developed Methods.* The novelty stems from the methods used to analyze the data. For example, our innovation may be to analyze pre-existing data in a faster, more efficient, accurate, precise, nuanced, or general way. This category also includes interdisciplinary work that introduces methods from one field to another. When the methods provide the novelty, we must emphasize why our approach is better. To do this we have to show that we understand what other researchers did and compare our results with theirs based on a meaningful metric.

Admittedly, this is a bit of a simplification, but most reports do fit into one of these two categories (and occasionally in both). Your approach in drafting the data description and methods depends on which category your work falls into. Either way, your goal is the same: convince your reader that your work is new and well conceived.

Chapter 3 describes how to write a data description. As mentioned there, the data description is a good place to start writing because it has a concrete goal, and the content should be mostly the same across different angles that our report could take.

Example: *DAWN Data Description*

The target publication for the DAWN analysis is an *MMWR* full report. According to *MMWR* guidelines, the methods section should primarily focus on the data-collection procedure because the novelty of the work comes from the dataset itself. The DAWN survey gives a more comprehensive look at drug-related emergency department visits. The methods section of an *MMWR* report about our DAWN findings might look like the following example.

The DAWN survey was collected from 2004–2011, but we focus on the most recent year's results in this report. The records come from a scientific sample of hospitals across the United States where hospitals in certain metropolitan areas of interest are oversampled. A hospital that is eligible to be sampled for the survey is a non-Federal, general hospital that has a 24-hour emergency department (ED). The data are weighted to account for the probability of the hospital being selected for the survey, nonresponse, and the total number of ED visits.

ED visits at each hospital are reviewed by medical staff who determine whether the emergency was related to drug use. If drug-related, the visit is categorized into one of eight case types: suicide attempt, seeking detox, under-age alcohol, adverse reaction, over-medication, malicious poisoning, accidental ingestion, and other abuse. The designation relies on the judgment of the medical staff, who use toxicology tests to determine the exact drugs responsible for the visit. The relative standard error of the proportion of ED visits for various case types is about 10%. Each record is also labeled whether or not it is a case of misuse or abuse. Illicit drug use is always considered drug abuse, but the nonmedical use of pharmaceuticals relies on the judgment of the medical staff and any additional information they receive during the visit. Information about alcohol usage is recorded if it is used in combination with another drug or if it is the sole drug in patients under 21.

These two paragraphs provide a description of the sampling process, the information collected for each case, and the precision obtained in the study. The entire first paragraph is reserved for how particular hospitals enter into the survey. These details are necessary to help support the claim that this dataset is representative and the most encompassing to date, allowing for questions to be answered that were previously unanswerable. ∎

Example: *Data Collection for Applying Benford's Law to Social Networks*

Golbeck's article, "Benford's Law Applies to Online Social Networks" appeared in *PLoS ONE* in 2015. The analysis brings well-known statistical methodology to bear on a new data problem that examines the size of social networks. The article describes the five sources of data that the author analyzes, some from others' work and some collected for the purpose of this analysis.

We analyzed data from five major social networking websites: Facebook, Twitter, Google Plus, Pinterest, and LiveJournal. ...We collected some of these datasets ourselves and used other datasets that had been created by others.

The author argues for the usefulness of their analysis by demonstrating that their approach can be applied to data collected by others. She also anticipates objections to their analytic approach for Twitter data, and below presents the problems that are being avoided with their alternative design for data collection.

Although there are existing Twitter social network datasets online, we collected our own data in this project in order to work with non-anonymized users so we could later analyze their account activity.

The author provides a detailed description of the selection of Twitter users to convince the reader of the quality of the data and appropriateness of the analysis. These claims are carefully stated, e.g., the distribution is described as *fairly uniform* rather than uniform. The use of *fairly* softens the claim so the reader knows that qualitatively, the data seem reasonable without forcing the author to test whether the data follow the uniform distribution. Chapters 8 and 9 discuss in more detail careful wording choices.   ∎

Example: *Data for the Tree-Ring Analysis*
In the Schofield et al. paper, the tree-ring data are described in terms of how the data are collected, what types of trees are measured, and where these trees grow.

Our tree-ring data are records of annual radial growth increments (i.e., ring widths) of living and well-preserved remnant (subfossil) Scots pine (Pinus sylvestris) growing near the latitudinal tree-line in Tornetrask, northern Sweden (Grudd et al. 2002; Briffa et al. 2008).

The citations allude to the fact that this is a commonly used dataset in the field. When explaining new methods, it can be useful to demonstrate the approach on a dataset that is established in the field; the reader is more likely to appreciate the gains made if the comparison is against something they are familiar with.

The authors describe what portion of the data they focus on and hint at why they analyze a subset of the dataset (they want the series of observations to be relatively long).

The full dataset consists of ring widths from 587 trees. We analyze a subset comprising $k = 247$ trees with at least 20 ring width observations after the year 1496, leading to a 500 year series.

Below, the authors call attention to the features of the data most relevant to their objective. Again, this assures the reader that this particular dataset is appropriate to compare proposed methods against those traditional in the field.

We use the Tornetrask data for two reasons. First, the correlation between the growth increments and climate is strong enough to make viable local reconstructions (see Section 4 for details). Second, the spread of age classes of the trees is relatively even through time. This makes the use of regional curve standardization (discussed in Section 3.2) viable for these data as well as traditional standardization (discussed in Section 3.1). ■

7.3.3 Methods of Analysis

Typically, we face many decision points when we analyze data, and we must describe our motivation and reasoning for the choices we made. We need to pre-emptively defend our work so that readers focus on the overall approach rather than wonder about individual choices. One way to defend our choices is to reference others who took a similar approach in similar scenarios. This strengthens our argument by appealing to precedent. However, we then need to distinguish our work from others' by emphasizing our unique contribution. Are we keeping all else equal, but assessing the effect of changing one aspect of the analysis? Are we adding to a baseline methodology? Are we generalizing another approach? Finding the balance between the status quo and novelty can be challenging.

If we choose to do something that is different from previously established methods, we want to anticipate someone asking "why" and show that we know others have done things differently, acknowledge that our approach is different, and explain why we intentionally took a different path. Perhaps we do not think the assumptions others have made are plausible, or we have found flaws in study designs. After we explain why we have deviated from others, we then must explain how our particular approach overcomes the issues we raised. A reader can more easily evaluate our work if we provide the context in which we developed our procedures and analysis.

Example: *Statistical Methods for Checking Benford's Law*
The bulk of the methods description in Golbeck's article is dedicated to data collection. This balance is appropriate for an analysis that applies traditional statistical methods to newly acquired data. However, the author does allot some space to describe the methodology because, while "standard", these methods had not previously been applied in this setting. They anticipate possible objections to their choices and present reasons as to why the proposed method of analysis is more appropriate than other, more traditional approaches used in the field. When we bring standard methods to a new context, it is important to explain any adjustments that were made to make the method applicable in the new context. For example, even though Benford's Law provides an exact distribution, in this context, the author merely uses it as a rough guide. The author defends her approach below.

Note that with datasets of this size, it is not appropriate to conduct a statistical hypothesis test for goodness of fit; over tens of thousands or millions of people, even a very tiny deviation would cause us to reject the null hypothesis.

The author also justifies her approach by referencing how others have used this same technique in similar situations in other fields.

Pearson correlations are a common way to measure how closely a distribution adheres to Benford's Law [20–22].

This claim reinforces the fact that the choice of technique is not arbitrary. ■

In general, you want to justify any modeling choice because a skeptical reader will ask why.

Example: *New Methods for the Tree-Ring Analysis*
The novelty of the work explained in Schofield et al. is the new methodology used to analyze tree-ring data for better understanding recent climate change. In their paper the authors analyze data that several other scientists have previously analyzed and compare their new approach to standard methodologies. By pointing out problems with existing procedures, they motivate their work. The main focus of one section in the paper is to clarify what the authors call "major problems" with each of the existing approaches. They explain how these issues can lead to "unintentional removal of climatic influences" and poor predictions.

TS has two major problems. First, the uncertainty about parameters in the standardization is ignored in all further modeling. Second, incorrectly modeled variation in the standardization output can distort modeling of the climate signal, inducing bias in the reconstructed climate values.

After they identify the problems, the authors explain how these problems propagate through the analysis and affect quantities of interest. By showing the consequences of these problems, they strengthen their case that a new approach is necessary.

The problem is that TS effectively detrends each tree-ring series in an attempt to remove the nonclimatic growth. This can easily lead to the unintentional removal of climatic influences on growth from the tree-ring chronology (Cook et al. 1995), that is, some of the low-frequency climate "baby" can get thrown out with the nonclimate "bathwater."

The authors also add strength to their argument by showing that other scientists have noted the same limitations. They also frame their new methodology as a more general methodology that encompasses the two standard approaches. This way the authors elevate their work beyond just another procedure to consider.

Instead of choosing between procedures, we prefer to consider a modeling framework that includes these two procedures as special cases. This provides flexibility to allow other assumptions that may be more appropriate for the observed data.

Above, the authors highlight the flexibility of their approach and show that by using it, they avoid making a model-selection decision that may introduce uncertainty.

When the authors make computational decisions that differ from the standard procedure, they explain their choices and compare the results with the original implementation, as described below.

This approach differs from that taken by Haslett et al. (2006) who did not use the full likelihood for inference. ...The main difference between the predictions appears to be the magnitude of warming/cooling. The actual time periods of relative warming and cooling are largely in common between all approaches.

By emphasizing that there is considerable agreement between the different computational approaches to the same model, the authors aim to convince the reader that any differences found later on between the standard procedures and the new one are not simply due to implementation differences. ∎

If you are a data scientist entering an applied field, it can be helpful to cite researchers within that field (not just data scientists and statisticians) to help anchor your argument for all readers.

7.3.4 Results

In the results section, we present the statistical results that are most relevant to our findings. Parsimony is key. If we present everything that we did, then we overwhelm the reader. Instead, we choose the details that matter i.e., that are essential to our argument. If there is a part of our analysis that is relevant but not part of our main findings, we summarize it in a few sentences and avoid going into detail. Similarly, we refer readers to an appendix for specific details rather than include them in the main text.

The results section should not be a chronological list of processing and analysis details. Let the storyboard narrative drive the organization and writing in the results section. That is, you want to describe what you found more than how you found it. Avoid "process" language, such as: "we ran a regression", "we made a bar chart to see if", "we replaced 999 values with NA", and "we conducted an analysis and found", and instead focus on the results with language such as: "a regression of X on Y shows", "in Figure 1 we see", "after removing missing values, we found", and "the regression and plots confirm".

Furthermore, the results section is not merely a place to list our findings but also a place to convince the reader that what we did is correct, worthy of study, and impactful. An important aspect of arguing for the correctness of our results is to avoid overselling them. Your writing should be straightforward about the generalizability of your work and avoid words that imply more certainty than you have. While the reader is learning about our findings we want them to be convinced that our involvement was necessary and our contribution was worthy of study. By making comparisons between our results and the results of others in previous studies, we emphasize what our work adds to the previous conversation. To show the impact of our results, we explain them in context, e.g., what are the practical implications of our work? If our findings end up being negative results, we take extra care to explain why these results still matter. For example, if our new method fails to improve the performance of the state of the art, we might emphasize the clarity of our conceptual model as the strength of our work.

Example: *DAWN Results*
We start off the results section of the DAWN report with a hint of the importance of the analysis, and we explain how we are able to make fair comparisons without going into too much detail about the procedural aspects.

The youngest (five and under) and the oldest (65 and over) together make up one in five of all drug-related visits to an emergency department (see the first row of Table 7.1). Moreover, the rate of visits among these children is four times the rate for children six to eleven years old (see the third row of Table 7.1). The rates make fair comparisons between sub-populations because they normalize the counts of drug-related emergency department visits per 100,000 people in an age group.

The next two paragraphs separately discuss sub-populations by looking at variables broken down by age: type of case, time of visit, and disposition (where the patient goes after the emergency department visit). In each paragraph, a figure is referenced and the major focus is described. The paragraphs are tied together by the examination of these three variables.

Figure 7.2 shows the breakdown of the type of case for each age group. Nearly all visits from those five and under are due to adverse reactions (70%) and accidental ingestion (26%). Furthermore, roughly one quarter of the cases of accidental ingestion for these children is due to the ingestion of pain medication, such as acetomenophine, aspirin, and ibuprofin, and half of the emergency visits for accidental ingestion occur after 6 pm.

The elderly visits are also not the result of abuse. An adverse reaction is the primary type of case (84%), with over-medication a distant second (8%). Together these types of cases make up nearly all drug-related ED visits for those 65 and over (see Figure 7.2). When we consider the disposition of the visits by the elderly (where the patient goes after the visit), we find an increase with age of patients being admitted to the hospital for additional inpatient care. Additionally, two in three of the visits from the elderly occur between 6 am and 6 pm.

Notice the parsimony in this section. There are many age groups that we do not discuss in the passage as well as types of drug-related ED visits that we do not delve into.

Although the figures show the full results, our text descriptions focus on the two age groups of interest based on our motivation to examine preventable emergency department visits for vulnerable sub-populations. ■

In addition to being parsimonious, it is important to connect your results to the wider body of work in the field. Do your results agree or disagree with other similar studies? If you do something different than is standard, acknowledge the difference and defend your choice. Be transparent about the aspects that didn't work; negative results are still results. By connecting your work to others you both prove that you understand the wider context of your own work and defend against those who might counter your results with "what about this other study?".

Example: *Results from the Social Network Analysis*
The first sentence of the results section of Goldberg's article states the findings, including both the positive and negative results. It reads as follows.

We found that the distribution of FSD [first significant digit] among friends in all five datasets closely followed the values expected from Benford's Law, with one interesting exception: the Pinterest following relationship.

A table of statistics is also provided to help compare approaches.

As mentioned earlier, the author, in defense of her statistical approach, calls upon the original work by Benford as a guiding principle rather than an exact law. This language (and the below statement) adjusts the expectations of the reader as to what is a considered a successful result.

Furthermore, conformance with Benford's Law has never been about a perfect statistical match to the predicted values—not even in Benford's original work on the subject [8, 9]. Rather, the relative frequencies of FSDs are the guiding principle.

The negative result pertaining to Pinterest data is construed as a useful insight, rather than a problem with the methodology.

This exposes an important point about applying Benford's Law: it can be violated when there is external influence over people's natural behavior. In the Pinterest case, we discovered the influence was an artifact of the system configuration. ■

Example: *Results of Model Checking in the Tree-Ring Analysis*
When comparing the results between new and old methods, Schofield et al. are transparent about their negative findings. Notice the careful use of *appears to be* and *are likely to* in the description of the results below. These phrases emphasize that the authors have obtained similar results but not exactly the same as others. These careful statements defend against a reader asking "If the results are the same, why do we need this fancier

model?". The authors also anticipate a reader's potential surprise at models appearing similar but resulting in noticeably different predictions.

There appears to be little difference between the various approaches in terms of predictive performance (Figure 5). All approaches are likely to overestimate when the true temperature value is below the observed mean and vice versa. There appears to be little difference between the models with the smooth mean for temperature and those with constant mean. This is somewhat surprising given the difference in predictions based on these models....all approaches (including those with scientifically implausible extremes) appear to be conservative at predicting values that are further from the observed mean.

Below, the authors translate their technical results into practical recommendations.

As there is no model clearly favored by the data and many modeling assumptions that could be considered (some additional assumptions are mentioned in Section 7), the results suggest: (i) modeling decisions matter, and (ii) the uncertainty we should have in the predictions is greater than that expressed by any one model.

These recommendations help less-technical readers.   ■

7.3.5 Figures and Captions

To choose figures for our article, we consider the storyline. What figures encapsulate the main take-away? You have probably made many graphs on your way through the analysis, but it is likely that none of these early figures are just right for publication. Now that you have a solid sense of what information you hope to convey in your report, it is easier to identify the visual comparisons needed to get your point across. Before we finalize the appearance of our graphs, we consider whether the graphs we have made consider adequately support our findings. We continue to make rough graphs in search of ones that best exemplify the core information. Ideally, we identify a single plot that features our main results and supplemental plots that support key points. After we have settled on the most effective plots to include in the article, we use the principles of visualization covered in Chapter 4 to improve their presentation.

Every figure needs a caption, a few sentences to accompany the figure. This text may seem repetitive because we also introduce the figure in the main body of our report. However, keep in mind that one strategy for quickly reading scientific papers is to read the beginning, the end, and look through the figures. If a reader looks at a graph in isolation, you want them to understand what is going on. An informative caption will make a graph capable of standing alone.

The caption should be self-contained and address the following.

- Describe what has been plotted.

- Summarize the conclusions drawn from the plot, i.e., what the reader should take away from the plot.

- Call attention to important details that a reader might not easily notice, especially if they are just glancing through the figures.

- Include any details about how the plot was produced that you do not want to clutter up the main text with.

Example: *DAWN Sample Figure Captions*
The *MMWR* full report is limited to at most three tables, figures, and boxes. After we review the plots and tables that we used in our storyboard, we selected two plots and one table to include in the report.

The first plot appears in Figure 7.2. We remade the figure that appears in the top right of our storyboard (Figure 7.1) by reordering the categories on the *x*-axis (which were originally in alphabetical). The new order reflects the prevalence of each type of case. The caption describes what is being plotted (visits by patient age and case type), points out the sub-populations that we want the reader to focus on (the youngest and oldest age), and conveys a take-away (the youngest and oldest age groups share adverse reaction as a major reason for drug-related ED visits, but differ in their secondary reason).

For the second plot, we modified the earlier plot on the bottom right of the storyboard that examines the relationship between age and disposition. In the new plot, which appears in Figure 7.3, we have focused on one disposition—admission to the hospital for further in-patient care. The caption begins by pointing out the upward trend with age. The graph is simple enough that an orientation to the axes and what is being plotted is unnecessary. However, the caption does clarify what is meant by *inpatient* by listing other types of hospital admission that are not included in this disposition.

Lastly, the caption for Table 7.1 first orients the reader to the values being displayed in each row, including a definition of the rates that appear in the bottom row of the table. The caption then focuses on the two groups of interest by noting their combined size, and goes on to draw attention to a compelling comparison between rates for the 5-and-under and the 6–11 age groups. ∎

7.4 The End

At the end of the paper we summarize our main findings, interpret what our results mean in context, discuss limitations, and point to where there is room for improvement or further analysis. This is your last chance to make an impression on your audience. Make sure that you have said all that you wanted to say.

Source	Age Group				
	5 & under	6–11	12–17	18–24	25–34
DAWN %	5.8	1.4	5.6	14.7	18.0
2010 Census %	7.9	8.0	8.3	9.8	13.3
Visits/100,000 people	1,201	288	1,107	2,455	2,217

	35 to 44	45 to 54	55 to 64	65-plus
DAWN %	14.3	15.3	10.2	14.8
2010 Census %	13.3	14.6	11.8	13.0
Visits/100,000 people	1,762	1,718	1,416	1,864

Table 7.1 Age Distribution for Drug-Related Emergency Department Visits. The top row of the table displays the percentage of drug-related visits to the emergency department by age group, and the second row gives the US population distribution from the 2010 Census. The third row provides rates of the number of visits per 100,000 persons. Together the youngest and oldest age groups account for 20.6% of all emergency department visits, which roughly matches their presence in the population (20.9%). Most noticeable is the rate of visits for the youngest is four times the rate for those aged 6–11 year.

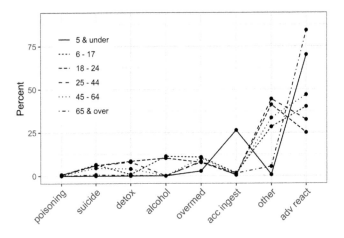

Figure 7.2 Case Type by Age. The distribution of drug-related emergency department visits is broken down by age and the type of case. A striking feature is the extreme percentages for adverse reactions in the youngest and oldest age groups. Those five and under suffer predominantly from adverse reactions, and secondarily from accidental ingestion. Those 65 and over also suffer primarily from adverse reactions, but over-medication is an important reason for visits. Neither of these two groups have noticeably high levels of cases of drug abuse.

7.4.1 Discussion

The discussion section takes a step back and gives our specific work a more general perspective. We want to broaden from the particular details of our findings to the field we are working in. After reading our discussion, a reader should know:

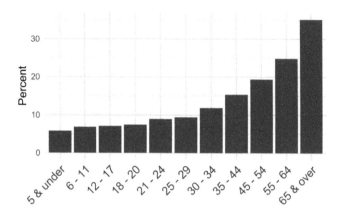

Figure 7.3 In-Patient Care After Visit. The percentage of drug-related emergency depart-ment visitors that are admitted to the hospital for in-patient care increases rapidly with age, for patients over 45. These percentages do not include those sent to intensive-care units, surgery, or psychiatric units.

- What are the key features of our analysis and results?

- Do our results confirm or contradict our initial expectations or earlier work?

- What problems did we run into, and if they still remain unsolved, what are our suggestions for starting to address them?

- Knowing what we know now, after doing our analysis, what recommendations do we have?

- What should someone work on next to build on our work?

Be careful of overstating your work. Be honest about the obstacles you faced and the generalizability of your findings. This does not weaken your report, but rather helps to ensure that your work is appropriately understood and used by others.

Example: *DAWN Discussion*
The discussion in our *MMWR* report begins with the limitations of the study so that the reader understands any problems with generalizability.

The DAWN data give a complete picture of drug-related ED visits across the United States by age, sex, race, and drug type that was previously inaccessible due to the lack of a comprehensive survey. However, there are some limitations that affect the results in this report. Information on race and ethnicity is often sparse, and some hospitals do not report this information at all since they consider this to be private information. Without race being consistently reported, we are not able to investigate how this factor interacts with ED visits. Another data limitation pertains to the variety of pharmaceuticals involved in the visits. Variety may be overstated because it can be challenging to determine whether or not a patient's current medications are related to the visit. This potential bias could be a greater problem for older patients since they often take multiple medications.

The next paragraph summarizes the findings for the youngest age groups, and makes suggestions for how to act on the findings with the goal of decreasing the number of drug-related ED visits.

We found adverse reactions and accidental ingestion to be the major types of drug-related ED visits for the youngest patients. Additionally, we found that visits for this group often occur during the evening when they are most likely home and under their parent's control. Emphasis on better storage of medication in households with small children has the potential to reduce drug-related ED visits by youths. In addition, better education for new parents about the dangers of medications in the household could possibly help decrease the number of incidents.

The end of the discussion summarizes the main features of drug-related visits for the elderly, provides a hypothesis of why they may occur, and offers suggestions for how to act on the findings to decrease the number of visits in this vulnerable age group.

Our investigation also found that visits from the elderly were commonly due to adverse reactions, and together with over-medication these cases make up nearly 95% of drug-related visits to the ED in this age group. The increase in inpatient services after the ED visit for the elderly may be attributable to the opportunity for a closer evaluation of the patient's ability to care for themselves or to deterioration of other health conditions brought on by the drug-related incident. For the elderly, clearer instructions and better monitoring for prescribed drugs (including interactions between multiple drugs) could help reduce drug-related ED visits and keep them out of in-patient care. ∎

Example: *Discussion of the Social Network Analysis*
Golbeck's discussion of her analysis begins with a simple claim about her findings.

We have shown that Benford's Law applies to relationships in online social networks.

The discussion continues with a justification of Benford's Law, based on examples of its application in other fields, and describes how using Benford's Law might help in identifying unusual observations for further review. How to interpret findings from a future analysis of new data is also discussed.

As is the case with forensic accounting investigations using Benford's Law, a deviation does not necessarily mean there is fraud happening. Given the large number of users on social media, it would be statistically unusual to have no accounts that naturally deviate from expected patterns; rather, deviation from a Benford distribution can flag accounts for additional review.

Given the kind of data that are examined, others may call into question the appropriateness of the analysis. The author anticipates this criticism, raises it as a serious concern, and warns others to exercise caution when applying the technique.

Truly random or representative sampling is difficult to do—and essentially impossible when connected components of a social network are important to the analysis. This raises the question as to whether the sample of accounts collected by a research team seriously deviates from normal patterns. While Benford's Law only addresses one aspect of expected behavior, major differences between a sample's FSD distribution and Benford's Law could indicate serious sampling problems.

The discussion includes further analysis and findings related to the appropriateness of the approach. The author explains that in many cases "structurally, the networks look like we would expect." The author mentions that she did uncover an example that "hints" at problems, but concludes that "further research is needed to understand the implications" of nonconforming data samples. ∎

Example: *Discussion of the Tree-Ring Analysis*
The discussion section of Schofield et al. serves many purposes. It first clarifies the core finding, as below.

Inferring historical climate from ring width data has many sources of uncertainty. Instead of trying to remove (or ignore) these [as in the previous approaches], we have built a model-based framework for including these sources. Care is needed when specifying alternative assumptions to ensure that the overall model continues to be identifiable.

The authors also point out limitations of their work and the need for ongoing research. One example is the following discussion of the impact of one of their assumptions.

Another assumption we have not varied is that the growth increments have a common response to temperature. Despite the trees existing in the same macro-environment, there may be fine scale local differences in the temperature, leading to a tree-specific response. Extending the models in this way is ongoing research and allows the inclusion of more realistic assumptions at the expense of several practical difficulties.

The authors anticipate criticism of the assumptions and aim to relax these assumptions in future research. ∎

7.4.2 Conclusion

The conclusion and discussion are not the same, yet they often appear in the same section. In particular, the conclusion should not introduce any new material. Whether in its own section or not, the conclusion should be brief and contain only a summary of the main points of the report. You may feel like you are repeating yourself when you write a conclusion, but that is the point. It is a good idea to remind readers what they should take away from our report after they have seen our analysis and sifted through some details. In addition, brief statement of our conclusions gives readers who are skimming a way to quickly understand what we covered in the middle sections that they may have skipped.

Example: *DAWN Sample Summary Box*
Although a typical *MMWR* publication does not have a separate conclusion, the summary box plays a similar role, describing the main take-aways of the report.

What is already known about this topic?
Non-compliance in prescribed medicine among the elderly and accidental ingestion by young children have been studied separately. For example, previous studies found pain relief medication to be the drug accidentally ingested by children the most often and poor memory and multiple physicians as reasons for over-medication among the elderly.

What is added by this report?
This report considers two vulnerable age groups. Because the groups are analyzed together, commonalities such as adverse reactions emerge. This report also considers the time of day the patients arrive to the emergency department and where they are sent after their visits.

What are the implications for public health practice?
After adverse reaction, the predominant reason for the young and old to have a drug-related emergency department visit are accidental ingestion and over-medication, respectively. Both problems can be addressed by closer supervision. Visits to the emergency department from these two vulnerable age groups tend to occur at different times of day, and they have different lengths of stays in the hospital system. This information can help hospitals with staffing and space allotment decisions. ■

Example: *Conclusions of the Social Network Analysis*
In Golbeck's paper, the discussion and conclusion appear together. The author makes two points in the concluding paragraph. One point claims the innovativeness of the approach for the setting, and the second point advocates for the importance of the tool. Both points are conveyed in the final sentence of the article.

The applicability of Benford's Law to social media is a new tool for analyzing user behavior, understanding when and why natural deviations may occur, and ultimately, detecting when abnormal forces are at work. ∎

Example: *Conclusions of the Tree-Ring Analysis*

The discussion section of Schofield et al. includes the conclusions as a subsection, which the authors have titled "Message for the Paleoclimate Community". That is, they phrase their conclusions as a message to the science community and place their findings in a larger framework with clear and important implications to researchers in this community. We paraphrase their message below.

We have demonstrated our methods are able to incorporate the assumptions of traditional approaches as special cases. Our approach allows us to relax assumptions long used out of necessity, giving flexibility to our model choices. Using the standard tree-ring data, we show how modeling choices matter. Alternative models fitting the data equally well can lead to substantially different predictions. These results do not necessarily mean any particular approach is wrong. However, if we are unsure whether the assumptions are correct and there are other assumptions equally plausible a priori, we will have unrecognized uncertainty in the predictions. We believe that such uncertainty should be acknowledged when using standardized data and default models.

The authors urge caution when interpreting default models since there is uncertainty that is often not accounted for, but they do not disparage the use of traditional approaches altogether. Statisticians often have the reputation of telling scientists what they cannot do with their data, but it is important to offer solutions to flaws found in others' analysis (as these authors have done) instead of just saying "no". ∎

7.5 The Beginning

The beginning of a report tends to be the most difficult to write. It is also one of the most important parts because after reading it, a reader decides whether or not to continue reading. The beginning should frame our work in terms of the context and previous work, discuss the results, and provide a road map for the report's presentation itself.

7.5.1 Background

The background section gives the reader context in terms of both the subject-area problem and the methodological problem. This section often includes a literature review that shows what has already been done with respect to the topic. When describing what has been done before, we want to frame others' work by pointing out what is still missing and alluding to how our own work fills the gap.

Ideally, before you started to work on your analysis, you read about what others have done in similar situations and where there are still open problems. The background section should highlight the reading you did that impacted the direction of your work, and not be an exhaustive list of every semi-relevant paper you consulted.

After reading the background, a reader should know:

- What are the key take-aways or approaches of the main studies or prominent authors in the field?

- What is missing in this literature?

- How does our work fit into this literature and/or fill in a gap?

Example: *DAWN Background*
The first sentence of the paragraph below introduces a broad goal and an existing challenge that hinders progress towards the goal. The next several sentences describe literature that studies the same sub-populations that interest us.

Health officials have tried to understand drug use in the United States, but are often limited by their study design. Other studies have taken advantage of the comprehensive DAWN data to answer questions about vulnerable populations. For example, a previous analysis of DAWN data found accidental ingestion to be a problem in the youngest age bracket. Researchers were able to look at an even more granular level to see how visits were distributed by age within the youngest age group and found that two year olds are most vulnerable. Among this age group, pain relievers were the most prevalent drug accidentally ingested followed by anxiety and insomnia drugs [37]. Improper ingestion of insomnia medication does not just affect the youngest age group. A previous study showed that about one third of the emergency department visits involving a popular treatment for insomnia resulted from over-medication [7]. In older adults, insomnia medications frequently are involved in cases where medicine and alcohol are ingested together. The negative effects of the interaction between alcohol and medications in general among older adults have also been studied and documented [29]. Our analysis extends the previous work and investigates new aspects of drug-related visits to the emergency room for the youngest and oldest age groups. We study when these visits occur and where the visitors are sent at the conclusion of their emergency visit.

The two final sentences explains how the current work fills in a gap. ■

Example: *Background for the Social Network Analysis*
The background information provided in Golbeck's paper dedicates two of its five paragraphs to a historical perspective on the formulation of Benford's Law. Presumably this is to provide authority to the approach the author has taken. The third paragraph presents Benford's Law and rather than providing intuition behind the law, the author states that the law has a *counterintuitive* appearance, but *Nevertheless, the law holds*

across many variations in measurement [10]. The remaining two paragraphs mention several applications of the law to temperatures, distances, and more modern cases related to the current application. Below is an excerpt.

> Benford's Law describes all these naturally occurring sets of numbers, and more. Specifically, some applications of Benford's Law are more relevant to our work. Benford often applies ...[12]. ...[13] ...[14]. Although no one has yet investigated how well Benford's Law describes social networks (online or offline) or social media, it has been shown to describe online human behavior through price distributions in eBay auctions [15].

The final statement shows that the work is a natural extension of preliminary findings in another study. This also implicitly justifies Benford's Law as a reasonable approach in the current study. ∎

Example: *Background for the Tree-Ring Analysis*
Schofield et al. does not contain a section called "Background" but does have a section called "Standard Methodology" which describes existing methods. Also, for those unfamiliar with the application area, the data section thoroughly explains the type of data that the standard approaches have used.

Since the journal in which this article appears has a statistical audience, the background includes information on the climate science. If the reader doesn't understand what the data look like and what methods have been used, they may have a difficult time understanding the full impact of the paper. If a similar article was published in a climate-science journal, the data and methodology used to analyze those data would be referred to more briefly with the understanding that the reader is familiar with the common approaches to the problem. ∎

When discussing the work of others it is important to appropriately cite the relevant papers, books, etc. where we obtained information. While we are reading, brainstorming, and drafting we track what references we are building off of so that we don't accidentally omit any citations in our final draft. In addition to acknowledging where we get our facts from, if we use an excerpt word-for-word, we need to explicitly quote this text. Failure to cite or mark quotes is plagiarism. This is a serious offense to academic integrity with major consequences.

Generally accepted, common-knowledge facts such as the number of inches in a foot, the team colors of a basketball team, or the singer of a song don't require a citation. However, when in doubt, add a citation. Beyond research that you read, it is also important to cite tools you use, such as computational packages or software that you take advantage of in your work. There are many style guides for how to cite, and before finalizing your writing, you should consult that of the publication you are aiming for. However, throughout the early draft process, you may find it simpler to just pick a style and use it consistently.

7.5.2 Introduction

The introduction usually goes in the opposite direction of the discussion section. Rather than move from the details of our study to the broader field, an introduction first gives an overview of the broader field and then identifies the specific question we are trying to answer within that space. When describing a statistical analysis, we must identify both the scientific and statistical question of interest and explain how our data and results provide answers to those questions.

After reading the introduction, a reader should know:

- What is the problem?

- What is the motivation?

- Why is it important?

- What was found?

You may think that you want to save your main conclusions for the end of the report, but you want to state them in the introduction as well. Telling the reader ahead of time what you found peaks their curiosity in how you obtained the results and may convince them to continue to read.

Typically, at the end of the introduction, the rest of the paper is outlined. A road map for the paper helps the reader know what to expect and helps them skip over sections to read what interests them.

Example: *DAWN Introduction*
The beginning of the introduction identifies the subgroups of interest and explains why we were previously unable to explore their drug-related emergency visit behavior.

We often think about the rise in drug use in the United States in terms of illegal drugs and the teenage to middle-aged populations, but understanding the types of drug use that impacts vulnerable age groups such as young children and the elderly is also an important consideration. However, assessing heterogeneity in drug-related emergencies across sex, race, age, etc. has been challenging due to a lack of comprehensive data.

The introduction continues by explaining what has changed so that we can now study drug-related visits more easily and completely.

To address this gap in knowledge, the United States Department of Health and Human Services consolidated data from the Drug Abuse and Warning Network (DAWN). These data contain information about drug-related visits to hospital emergency departments (EDs) in over 250 hospitals across the country.

The introduction ends with a more specific description of the sub-population of interest and an overview of the major findings.

We focus on the youngest (five and under) and oldest (65 and over) age groups because to some extent, their drug-related ED visits could be prevented by increased parental or medical supervision. Combined, these two groups make up more than one in five drug-related visits to emergency departments. For the youngest children, visits are due to an adverse reaction or accidental ingestion, and visits for children five and under are over-represented when compared to the 6 to 11 year olds. The elderly population's drug-related ED visits are primarily due to an adverse reaction or over medication, and the proportion of these patients returning home directly after the ED steadily declines with age. ∎

Example: *Introduction for the Benford Law Application*
The introduction to Golbeck's paper mirrors the paper's organization. The introduction starts with "Benford's Law states" and provides the formula. Then, the background material related to other applications of Benford's Law is briefly described. Finally, the author emphasizes that even though Benford's Law has been used in many other scenarios, she is the first to apply it to social networks.

Benford's Law is frequently used in forensic accounting, where a distribution of first digits that is outside the expected distribution may indicate fraud [1]. Research has also shown that it applies to genome data [2], scientific regression coefficients [3], election data [4, 5], the stock market [6], and even to JPEG compression [7]....To our knowledge, this is the first time Benford's Law has been applied to social networks.

The remainder of the introduction summarizes what aspects of social networks follow Benford's Law and how we can use deviations from Benford's Law to diagnose "suspicious activity" or "malicious or irregular behavior". The author is transparent about both the positive and negative results. ∎

Example: *Introduction for the Tree-Ring Analysis*
When introducing their work, Schofield et al. begin by setting the stage with a "scientific consensus". They then bring focus on a particular practical problem.

Instrumental meteorological records from the past 200 years support a scientific consensus that climate is changing (Trenberth et al. 2007). What is less clear is how to interpret these changes in a historical context spanning many hundreds, or thousands of years.

The authors go on to explain the statistical aspect of the problem and the current solution present in the literature.

The statistical problem is to infer historical climate without having direct observations beyond the most recent two centuries. A common solution is to index historical climate values using proxy observations from natural archives such as tree rings, lake sediments, bore holes, corals, and ice cores (see, e.g., Jones and Mann 2004; Jansen et al. 2007).

The second paragraph of the introduction refers to many studies in the field conducted by others. Then, the third paragraph describes a problematic aspect of these approaches.

One consequence is that uncertainties from Steps (i) and (ii) above do not propagate through to the final predictions.

After pointing out the problems in other work, the authors give a high-level description of their solution.

To overcome these problems, we merge the three steps... Simultaneously describing climatic and nonclimatic influences on the raw data makes it possible to overcome difficulties encountered in multi-step procedures.

They wrap up the introduction with statements about the benefits of their new method.

In our framework, the different approaches can be viewed as differences in modeling assumptions and comparison between procedures becomes a choice between models. Once a modeling framework has been specified, we can consider alternative assumptions and examine how the predictions of historical climate depend on changes in both the standardization and reconstruction components of the model. If the predictions change substantially, this shows sensitivity of the modeling assumptions and can help reveal limitations of the data. ∎

7.5.3 Abstract

An abstract is often left for last. Having written the rest of the paper, it can be easier to summarize our work in a compressed format. Journals have varying requirements for the word length of abstracts, but between 100 and 500 words is typical. With the tight word limit, conciseness is key. The abstract should be able to stand alone, and the approach and results should be revealed in a simple way. However, repetition between the abstract and the introduction is common. Every abstract should include four main types of information:

- the main objective and rationale of our project,
- an outline of methods used,
- results of the project, and
- conclusions about the implications of the project.

It can be helpful to read a few abstracts of papers in your field to get a sense for what these look like. Think of the abstract as an advertisement for your work. You want to "sell" your paper. When deciding which papers to read, a reader goes through many abstracts before choosing what they think will be most relevant to their interests.

Example: JASA *Instructions for Abstracts*

JASA provides specific instructions for abstract writing.

Each manuscript should contain an extended abstract of 200 words. The first 100 words should succinctly describe the paper's motivation and contribution. For the benefit of JASA's broad readership, the remainder of the abstract should amplify and illustrate, preferably using concrete examples and interesting special cases. Do not cite references in the abstract. ∎

Example: *DAWN Abstract*

The abstract below explains the goal of our study (find patterns associated with substance abuse) and how we achieve this goal (using drug-related visits to the hospital emergency departments from DAWN). Then it explains our main findings. From the abstract, the reader knows that our focus is the very young and the elderly.

By using drug-related visits to the hospital emergency departments from the Drug Abuse and Warning Network we uncover patterns associated with adverse reaction, accidental ingestion, and over medication across age. Nearly all drug-related emergency hospital visits for children five and under are due to an adverse reaction or accidental ingestion. These young children are over-represented in drug-related visits when compared to 6 to 11 year olds. The elderly population's drug-related emergency visits are primarily due to an adverse reaction and over medication, and patient admission to the hospital following the visit steadily increases with age. ∎

7.5.4 Title

After we have the heart of our report written, we start thinking about a title. We want the title to be informative, searchable, and memorable. To help you come up with a title, write one sentence that states what you did and another sentence that states what you found. From these two sentences, you can put a title together that has the form "We Did X: We Found Y". Then, you can work on making the title less formulaic.

The title can help focus the message of our report. If we realize our work no longer matches our title, then we decide whether we need to change our writing focus or choose a new title that reflects the shift in our story.

Example: *DAWN Title*

The first part of the title below explains what we are exploring, and the second part hints at what we found.

Drug-Related Emergency Department Visits:
The young and the elderly in need ∎

7.5.5 Keywords

Keywords help a reader find the most relevant research articles to read. Often journals give some guidelines about selecting keywords, but common advice includes avoiding words that appear in the title and choosing words that refer to specific techniques, methodology, and/or topic area of application. It can be helpful to look at the keywords in the articles that are most relevant to your work and are referenced in your manuscript. You can search online for a preliminary set of keywords to see if you find papers similar to yours. If you find too many papers using your keywords, try to be more specific.

Example: *DAWN Keywords*
We chose the following keywords by thinking about someone who might search for our paper. What search terms would they use? We want to make sure our keywords line up so that they are more likely to find our work.

drug abuse, aging, DAWN, accidental ingestion, over medication ■

7.6 Notes

Strategies from this chapter include the "They say/I say" approach to writing which was originally explained in Graff and Birkensten's book of the same name [18]. We recommend this book as supplementary reading. When deciding on the order to write the first draft in, Andrew Gelman's blog post was consulted [14]. For more general advice about making time for writing see Silvia [35] and Chapter 17 of Alley [1].

The guidelines for the three journals discussed in this chapter can be found at the Centers for Disease Control and Prevention [8], *PLoS ONE* [31], and *JASA* [24].

The papers used throughout as examples were the tree-ring paper by Schofield et al. [33] and the Benford's Law paper by Golbeck [17]. The citations made in the "Background for the Application of Benford's Law" example refer to (in order): Pietronero et al. [30], Barabasi and Albert [3], Asur et al. [2], and Giles [16]. The citations made in the "Introduction for the Benford Law Application" example refer to (in order): Hoyle et al. [20], Diekmann [11], Tam and Gaines [39], Roukema [32], Hill [19], and Fu, Shi, and Su [13].

Example figures for Activity 7.8.3 on writing captions are inspired by (in order): Tran, Wiskow, and Aziz [40], Gibbs and Goossens [15], Kiefer [27], and Stoudt [36].

Activity 7.8.5 on a journal's scope was inspired by the "British Ecological Society Workshop: How to be a Good Reviewer" at the International Statistical Ecology Conference in 2018 attended by Sara Stoudt. Examples of the scope information for journals can be found at *Biometrika* [5], *JASA* [23], *JSE* [22], and *Environmetrics* [12]. The abstracts used in this activity can be found here (in order): Huang et al. [21], Crawford, Wu, and Heimer [9], Kent, Ganeiber, and Mardia [26], Kasprowicz and Musumeci [25], Bogomolov and Heller [6].

Activities 7.8.8, 7.8.9, and 7.8.10 include sample blog titles, introductions, and conclusions that come from student work, with the exception of the last two introductions which come from Silge [34] and Bastia [4]. The press-release titles in these activities come from UC Berkeley Media Relations [41], Maclay [28], University of Leicester Press Office [42], and De Witte [10], and the sample press release conclusions come from Sweeny [38] and UCB Public Affairs [41].

For more practice, consider working on Portfolio Pieces E, H, or U.

7.7 References

[1] M. Alley. *The Craft of Scientific Writing.* Springer-Verlag, New York, 3rd edition, 1996.

[2] S. Asur, B. A. Huberman, G. Szabo, and C. Wang. Trends in social media: Persistence and decay. *Proceedings of the Fifth International AAAI Conference on Weblogs and Social Media,* 2011.

[3] A. L. Barabasi and R. Albert. Emergence of scaling in random networks. *Science,* 286(5439):509–512, 1999.

[4] H. Bastia. In a lather over triclosan? thumbs down to fear mongering soap operas. https://blogs.scientificamerican.com/absolutely-maybe/in-a-lather-over-triclosan-thumbs-down-to-fear-mongering-soap-operas/, November 2014.

[5] Biometrika. About the journal. https://academic.oup.com/biomet/pages/About.

[6] M. Bogomolov and R. Heller. Assessing replicability of findings across two studies of multiple features. *Biometrika,* 105(3):505–516, 2018.

[7] D. M. Bush. Emergency department visits attributed to overmedication that involved the insomnia medication zolpidem. *The Center for Behavioral Health Statistics and Quality Report,* 2014.

[8] Centers for Disease Control and Prevention. *Morbidity and Mortality Weekly Report (MMWR) Weekly: Instructions for Authors.* https://www.cdc.gov/mmwr/author_guide.html, January 2019.

[9] F. W. Crawford, J. Wu, and R. Heimer. Hidden population size estimation from respondent-driven sampling: A network approach. *Journal of the American Statistical Association,* 113(522):755–766, 2015.

[10] M. De Witte. A doctor's reassurance speeds healing from an allergic reaction, find Stanford psychologists. Stanford News Service, September 2018. https://news.stanford.edu/press-releases/2018/09/04/doctors-reassurance-speeds-healing/.

[11] A. Diekmann. Not the first digit! Using Benford's Law to detect fraudulent scientific data. *Journal of Applied Statistics,* 34(3):321–329, 2007.

[12] Environmetrics. Overview. *Environmentrics.* https://onlinelibrary.wiley.com/page/journal/1099095x/homepage/productinformation.html.

[13] D. Fu, Y. Q. Shi, and W. Su. A generalized Benford's law for JPEG coefficients and its applications in image forensics. *Electronic Imaging, International Society for Optics and Photonics,* 2007.

[14] A. Gelman. Advice on writing research articles. https://statmodeling.stat.columbia.edu/2009/07/30/advice_on_writi/, April 2018.

[15] A. L. Gibbs and E. T. Goossens. The evidence for efficacy of HPV vaccines: Investigations in categorical data analysis. *Journal of Statistics Education*, 21(3), 2013.

[16] D. E. Giles. Benford's law and naturally occurring prices in certain eBay auctions. *Applied Economics Letters*, 14(3):157–161, 2007.

[17] J. Golbeck. Benford's law applies to online social networks. *PLoS ONE*, 10(8), 2015.

[18] G. Graff and C. Birkenstein. *They Say / I Say: The Moves That Matter in Academic Writing*. W. W. Norton & Company, New York, 2009.

[19] T. P. Hill. The first digit phenomenon: A century-old observation about an unexpected pattern in many numerical tables applies to the stock market, census statistics and accounting data. *American Scientist*, 86(4):358–363, 1998.

[20] D. C. Hoyle, M. Rattray, R. Jupp, and A. Brass. Making sense of microarray data distributions. *Bioinformatics*, 18(4):576–585, 2002.

[21] X. Huang, I. R. Iliev, V. Lyubchich, and Y. R. Gel. Riding down the Bay: Space-time clustering of ecological trends. *Environmetrics*, 29(5-6), 2017.

[22] Journal of Statistics Education. Guidelines for JSE authors. *Journal of Statistics Education (JSE)*. http://jse.amstat.org/jse_author_info.htm.

[23] Journal of the American Statistical Association. Aims and scope. *Journal of the American Statistical Association (JASA*. https://www.tandfonline.com/action/journalInformation?show=aimsScope&journalCode=uasa20.

[24] Journal of the American Statistical Association. Instructions for authors. *Journal of the American Statistical Association (JASA)*. amstat.tandfonline.com/action/authorSubmission?show=instructions&journalCode=uasa20.

[25] T. Kasprowicz and J. Musumeci. Teaching students not to dismiss the outermost observations in regressions. *Journal of Statistics Education*, 23(3), 2015.

[26] J. T. Kent, A. M. Ganeiber, and K. V. Mardia. A new unified approach for the simulation of a wide class of directional distributions. *Journal of Computational and Graphical Statistics*, 27(2):291–301, 2016.

[27] L. Kiefer. State unemployment rate dataviz: An update to a remix of a remix. http://lenkiefer.com/2020/05/22/state-unemployment-rate-dataviz/, May 2020.

[28] K. Maclay. Study asks why students with more to gain from charter schools are less likely to apply. Media Relations, UC Berkeley, April 2018. http://news.berkeley.edu/2018/04/03/study-asks-why-students-with-more-to-gain-from-charter-schools-are-less-likely-to-apply/.

[29] M. Mohanty and P. Slattum. Alcohol, medications, and older adults. *Age in Action*, 26(3):1–5, 2011.

[30] L. Pietronero, E. Tosatti, V. Tosatti, and A. Vespignani. Explaining the uneven distribution of numbers in nature: the laws of Benford and Zipf. *Physica A: Statistical Mechanics and its Applications*, 293(1):297–304, 2001.

[31] Public Library of Science. Submission guidelines. PLoS ONE. https://journals.plos.org/plosone/s/submission-guidelines.

[32] B. F. Roukema. Benford's Law anomalies in the 2009 Iranian presidential election. *arXiv*, 2013.

[33] M. R. Schofield, R. J. Barker, A. Gelman, E. R. Cook, and K. R. Briffa. A model-based approach to climate reconstruction using tree-ring data. *Journal of the American Statistical Association*, 111(513):93–106, 2016.

[34] J. Silge. You must allow me to tell you how ardently I admire and love natural language processing. https://juliasilge.com/blog/you-must-allow-me/, March 2016.

[35] P. J. Silvia. *How to Write a Lot*. APA LifeTools, 2nd edition, 2018.

[36] S. Stoudt. Geostatistical models for the spatial distribution of uranium in the continental United States. In *Advances in Geocomputation*. Springer, New York, 2017.

[37] Substance Abuse and Mental Health Services Administration (SAMHSA), Office of Applied Studies. *Emergency Department Visits Involving Accidental Ingestion of Drugs by Children Aged 5 or Younger*. Technical report, National Criminal Justice Reference Service, 2010.

[38] C. Sweeny. Religious upbringing linked to better health and well-being during early adulthood. Harvard T. H. Chan School of Public Health, September 2018. https://www.hsph.harvard.edu/news/press-releases/religious-upbringing-adult-health/.

[39] C. W. K. Tam and B. J. Gaines. Breaking the (Benford) law: Statistical fraud detection in campaign finance. *The American Statistician*, 61(3):218–223, 2007.

[40] T. Tran, C. Wiskow, and M. A. Aziz. Skewed and flexible skewed distributions: A modern look at the distribution of BMI. *American Journal of Undergraduate Research*, 14(2), 2014.

[41] UCB Public Affairs. Correctional officers at high risk for depression, PTSD, suicide, survey finds. Media Relations, UC Berkeley, April 2018. http://news.berkeley.edu/2018/08/23/california-correctional-officers-at-high-risk-for-depression-ptsd-and-suicide-new-survey-finds/.

[42] University of Leicester Press Office. Juices from damaged salad leaves massively stimulate *salmonella* growth and salad leaf colonisation, study shows. University of Leicester Press Office, November 2016. https://www2.le.ac.uk/offices/press/press-releases/2016/november/juices-from-damaged-salad-leaves-massively-stimulate-salmonella-growth-and-salad-leaf-colonisation-study-shows.

7.8 Activities

7.8.1 Keep a Brainstorm Diary

Before you can write a first draft, you need some material to write about. You may have some work in mind, but if not, it can be helpful to capture the passing thoughts you have and formalize them later. As you go through your day, if something sparks an idea, write it down. It doesn't have to be a fully formed thought. It can be just something that peaked your curiosity or made you wonder. After about a week, chat about your diary entries with a peer to help you flesh out a more concrete analysis idea to write about.

Here are some sample diary entries.

I read the article about Google Base data has 50 million unique drawings of circles, which is designed to investigate different language calligraphy shapes people's habits differently. Kind of interesting, mine is a clockwise circle, typical for Chinese based handwriting.

The library is so crowded all the time. How does Google use insights to predict how crowded a cafe or restaurant will be at a certain time and how can we apply that to seats in a library? Nothing more frustrating than going on to campus and having a hard time finding a place to work.

We are currently trying to book a Spring Break vacation for 6 people, including myself. Selecting the destination was a massive pain to begin with—everyone in our group had wildly different ideas and budgets. After having settled on a location, finding time to satisfy the requirements, budgets, wants and schedules of every one accompanying us is turning out to be somewhat like an incredibly complex optimization problem.

I have a Google Home speaker, and more recently a nice set of Phillips Hue lightbulbs as a Christmas present. My house is slowly becoming "smart". I can't help but imagine how "smart" my house will be in 30 years. How many new internet connected products will my house be made of? In which ways will my appliances talk to one another? I remember my grandma telling me stories of her childhood without plumbing or electricity. She witnessed so much technological advancement in her lifetime, and I know I will in mine.

7.8.2 Writing a First Draft

The best way to learn to write is to practice, and to practice you need a data analysis to write about. You have a few options. You can formalize something that comes out of your brainstorm diary above. You can find a dataset you are curious about and explore it using Exploratory Data Analysis (EDA) or some simple analysis. The keyword here is "simple". Do not get bogged down in fancy statistical methods at this point; your focus is on writing. You can also write about an analysis that you are doing for a class, for work, or just in your free time. If you already have a question, collected data, and have done some analysis, then you can turn to communicating your results. Alternatively, to lower the barrier to entry, we have provided some preliminary analyses that you can practice writing about.

Write a first draft about the EDA of all commercial flights departing from the San Francisco International Airport in 2008 (found at the end of Section 7.8.12). Use the figures and tables there for inspiration, they do not all need to be included in the draft.

Write the following sections in the order listed. You need only write a paragraph or two for each section.

- *Make a storyboard to develop your argument and story.* Which figures and tables are needed to explain your findings?

- *Use the storyboard to draft your conclusions.* Write a bullet-point outline summarizing the figures and tables you chose.

- *Write the middle of the report.* Include descriptions of the data and the methods used in the analysis with the emphasis appropriate for the category of analysis (new data, traditional methods).

- *Refer back to your storyboard summary to help you write about your findings and analysis.* Can you easily find any other studies of flight delays (or travel delays more generally) that you could connect your work to?

- *Write your discussion.* What are the limitations of your results? What could others study in the future related to your findings?

- *Rewrite your conclusions.* This should be a reiteration of your findings that occur elsewhere in your report. Make sure a reader knows the major take-aways.

- *Write your introduction.* Make sure you provide context, summarize your main findings, and emphasize why they are important.

- *Choose a title.* Choose an informative yet possibly clunky title and then refine it to be catchier.

7.8.3 Practice Writing Captions

Write a caption for each of Figures 7.4, 7.5, and 7.6, and Table 7.2. Each has a hint on what you might write about in the caption.

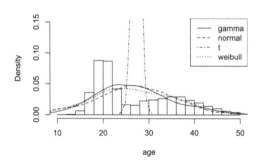

Figure 7.4 Hint: which fit wins?

Study	A	B	C
Control: Infected	14	107	321
Control: Total	400	800	6000
Vaccine: Infected	1	6	27
Vaccine: Total	450	850	6200

Table 7.2 Hint: what numbers or proportions should I focus on? Tables need captions too.

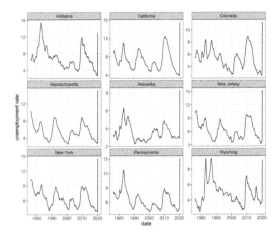

Figure 7.5 Hint: why might someone NOT standardize the y-axis scale in this case?

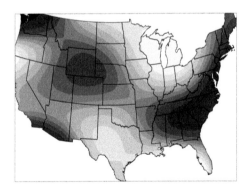

Figure 7.6 Hint: a darker color represents larger amounts of trace uranium. What natural features line up with the dark pockets?

7.8.4 Rewriting a Caption

Figure 7.7 shows conditional coefficients for day, longitude, and latitude, in the presence of one another in a model relating space and time covariates to average daily temperature. The caption for Figure 7.7 explains what to look for in the plot, but it fails to provide context for how the three figures relate to one another and give a concrete answer to whether or not the interaction terms are significant. Context is a key aspect of what makes a good caption. Rewrite the caption to include an explanation of what each plot is telling us about how interaction terms behave in this example. In your rewrite consider the following:

- What are the overall trends in relationships between space and time?

- What do the uncertainty bands tell us about what is going on?

- Call attention to at least one feature in each plot and explain what it says about the interaction term.

- Provide an overall conclusion that the reader should take away from all three plots interpreted jointly.

7.8.5 Scope

An important part of the writing process is choosing where to publish your writing since this will impact how you write about your work. This activity aims to help you better understand differences between journals. We provide an overview of the scopes of a few different journals and a few abstracts from these journals. Use the content and style of the abstracts, to choose which abstract is in which journal. Note the aspects of each abstract that convinced you of the matching journal.

Scope

1. "*Biometrika* is primarily a journal of statistics in which emphasis is placed on papers containing original theoretical contributions of direct or potential value in applications. From time to time, papers in bordering fields are also published."

2. "Established in 1888 and published quarterly in March, June, September, and December, the *Journal of the American Statistical Association (JASA)* has long been considered the premier journal of statistical science. Articles focus on statistical applications, theory, and methods in economic, social, physical, engineering, and health sciences. Important books contributing to statistical advancement are reviewed in *JASA*"

3. "The Journal of Statistics Education disseminates knowledge for the improvement of statistics education at all levels, including elementary, secondary, post-secondary, post-graduate, continuing, and workplace education. The journal is

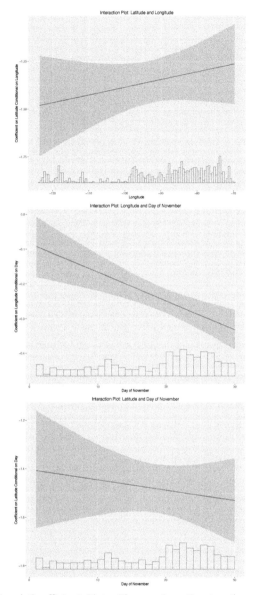

Figure 7.7 Conditional Coefficient Plots. The y-axis estimates the coefficient of one variable on average daily temperature conditional on the other variable. An uncertainty band is included, and the histogram portion of the plots shows the spread of the actual data. If there was a significant interaction term, we would see a change in slope of the curve in different regions of the domain.

distributed electronically and, in accord with its broad focus, publishes articles that enhance the exchange of a diversity of interesting and useful information among educators, practitioners, and researchers around the world. The intended audience includes anyone who teaches statistics, as well as those interested in research on statistical and probabilistic reasoning. Submissions are rigorously refereed using a double-blind peer review process.

Manuscripts submitted to the journal should be relevant to the mission of *JSE*. Possible topics for manuscripts include, but are not restricted to: curricular reform in statistics, the use of cooperative learning and projects, innovative methods of instruction, assessment, and research (including case studies) on students' understanding of probability and statistics, research on the teaching of statistics, attitudes and beliefs about statistics, creative and tested ideas (including experiments and demonstrations) for teaching probability and statistics topics, the use of computers and other media in teaching, statistical literacy, and distance education. Articles that provide a scholarly overview of the literature on a particular topic are also of interest. Reviews of software, books, and other teaching materials will also be considered, provided these reviews describe actual experiences using the materials."

4. "The *Journal of Computational and Graphical Statistics* (*JCGS*) presents the very latest techniques on improving and extending the use of computational and graphical methods in statistics and data analysis. Established in 1992, this journal contains cutting-edge research, data, surveys, and more on numerical graphical displays and methods, and perception. Articles are written for readers who have a strong background in statistics but are not necessarily experts in computing."

5. "*Environmetrics*, the official journal of The International Environmetrics Society (TIES), an Association of the International Statistical Institute, is devoted to the dissemination of high-quality statistical and mathematical research in the environmental sciences. Articles must proffer substantive, new statistical or mathematical advances, answer important scientific questions in the environmental sciences, or must develop novel or enhanced statistical methodology with clear applications to environmental science. New methods should be illustrated with recent environmental data."

Abstract

1. "The Chesapeake Bay Program, initiated in 1983, is a regional partnership between several state governments, federal agencies, and advisory groups that is involved in the cleanup and restoration of the Bay. To study the ecological trends in the area, we propose a new data-driven procedure for optimal selection of tuning parameters in dynamic clustering algorithms, using the notion of a stability probe. We refer to the new procedure as Downhill Riding (DR) because of the dynamics of the clustering stability probe. We study the finite sample performance of DR when clustering benchmark Iris data and synthetic times

series, and illustrate the methods using data on water quality in the Chesapeake Bay. "

2. "Estimating the size of stigmatized, hidden, or hard-to-reach populations is a major problem in epidemiology, demography, and public health research. Capture-recapture and multiplier methods are standard tools for inference of hidden population sizes, but they require random sampling of target population members, which is rarely possible. Respondent-driven sampling (RDS) is a survey method for hidden populations that relies on social link tracing. The RDS recruitment process is designed to spread through the social network connecting members of the target population. In this article, we show how to use network data revealed by RDS to estimate hidden population size. The key insight is that the recruitment chain, timing of recruitments, and network degrees of recruited subjects provide information about the number of individuals belonging to the target population who are not yet in the sample. We use a computationally efficient Bayesian method to integrate over the missing edges in the subgraph of recruited individuals. We validate the method using simulated data and apply the technique to estimate the number of people who inject drugs in St. Petersburg, Russia. Supplementary materials for this article are available online."

3. "The need for effective simulation methods for directional distributions has grown as they have become components in more sophisticated statistical models. A new acceptance-rejection method is proposed and investigated for the Bingham distribution on the sphere using the angular central Gaussian distribution as an envelope. It is shown that the proposed method has high efficiency and is also straightforward to use. Next, the simulation method is extended to the Fisher and Fisher-Bingham distributions on spheres and related manifolds. Together, these results provide a widely applicable and efficient methodology to simulate many of the standard models in directional data analysis. An R package simdd, available in the online supplementary material, implements these simulation methods."

4. "One econometric rule of thumb is that greater dispersion in observations of the independent variable improves estimates of regression coefficients and therefore produces better results, i.e., lower standard errors of the estimates. Nevertheless, students often seem to mistrust precisely the observations that contribute the most to this greater dispersion. This paper offers an assignment to help students discover for themselves the value of the observations that are farthest from the mean."

5. "Replicability analysis aims to identify the overlapping signals across independent studies that examine the same features. For this purpose we develop hypothesis testing procedures that first select the promising features from each of two studies separately. Only those features selected in both studies are then tested. The proposed procedures have theoretical guarantees regarding their control of the familywise error rate or false discovery rate on the replicability claims. They

can also be used for signal discovery in each study separately, with the desired error control. Their power for detecting truly replicable findings is compared to alternatives. We illustrate the procedures on behavioural genetics data."

See Section 7.6 to find the true sources and check your guesses.

7.8.6 Space

Along with choosing the right place to publish your writing in terms of content and style, you also must consider different space constraints. The way we approach a five-page paper is very different from the way we approach a 20-page paper.

In this activity you explore the typical length of articles in two journals and look through the latest issue of each to map out the articles, noting how many pages the authors designate for each section. This exercise gives you some guidance on how to pace your own work under a variety of space constraints.

- Shorter form journal: *Proceedings of the National Academy of Sciences of the United States of America* (*PNAS*)

- Medium/long form journal: *Journal of the American Statistical Association* (*JASA*)

For each of the journals listed above, go to the latest online edition. Select at least three articles in each journal, go through each, and record how many pages and paragraphs are devoted to the following sections: introduction, background/literature review, methodology and analysis, discussion, and conclusion.

7.8.7 Coming up with Titles

For each of the following titles we provide some hints about how to rewrite them to be more precise and eliminate unnecessary words.

Original Title Would Catching More Wi-Fi Signals Help to Estimate the Locations of Devices?
Suggestion Can we be more clear? What does it mean to "catch more" signal?

Original Title: Probability and the Real World: Minimize Taking Delayed Flights
Suggestion: Are there any extraneous fluff words that we can remove?

Original Title: Internet Defense: Understanding the Trends in Spam
Suggestion: For context, each row in the dataset inspiring this example includes characteristics of an e-mail and a label of whether the e-mail is "spam" or "not spam". Can we narrow the scope emphasized in the title?

7.8.8 Titles for Broader Publics

Unlike in a report where we often write the title as one of the last steps in the writing process, it can be helpful to come up with a title before we write an informal piece, such as a blog post. The title can help us keep the writing focused. Also unlike a report, you want a title that draws readers in but does not necessarily tell them the whole story right away. Maintain some mystery. Think about what titles make you want to click on and read, but avoid "clickbait" strategies that do not deliver on their promises.

Consider a question or command as a title. Use the following examples to revise the titles from the previous activity using these strategies.

Blog Titles That Question
If social media drives crime, can we mine it to prevent crime?
Are Plots Enough?
Does music make us less productive?

Commanding Blog Titles
Expectant Mothers: Quit Smoking for your Newborns!
Look, (yes, actually look), at the data

When reaching a broader audience with the help of a journalist, we often come up with the title after the press release has been written. A catchy title that draws attention to the story is a good model to follow. A subtitle that is more informative makes a useful accompaniment but is by no means necessary. Pick two titles from the previous activity and update them in the context of a press release using the following examples as inspiration.

Simple Press-Release Titles
Correctional officers at high risk for depression, PTSD, suicide, survey finds
Study asks why students with more to gain from charter schools are less likely to apply
Religious upbringing linked to better health and well-being during early adulthood

The title:subtitle Template
Juices from damaged salad leaves massively stimulate Salmonella growth and salad leaf colonisation, study shows
University of Leicester team show leached juices from leafy vegetables enhance growth and virulence of food poisoning bug

7.8.9 Introductions for Broader Publics

When writing for a broader audience we want to quickly grab our reader's attention. In the draft stage it can be helpful to write two introductions from different angles. Since an informal piece, like a blog post, is typically much shorter (300–600 words) than a formal piece of writing, it is not as time-intensive to keep options open. Two introductions give some flexibility in determining which turns into the most engaging storyline. When choosing which of your proposed angles to further pursue, put yourself in your reader's shoes. What would you want to know? What would keep you reading?

Consider the following examples of strategies used in blog post introductions. Adapt your introduction from Section 7.8.2 to make it more appealing to a broader audience. Feel free to decrease the formality of the language.

Questions in Blog Post Introductions The two introductions below draw the reader into the blog post through compelling questions.

You are probably laughing at Donald Trump's language style: "Great people." "Believe me." "Not good." But is he really special, or are the recent presidents all shifting to a simpler language style in general? Let's analyze the State of the Union Addresses and find out!

Did you turn on the radio today or hop on Spotify? Almost everyone, including you, has probably listened to a song today. Listening to music has become an everyday part of our lives from listening to music while working or jamming out at a party! This brings up questions like when do we listen to music? When should we listen to music? And how should we listen to music?

Attention-Catching Blog Post Introductions The following two introductions use allusions to common phrases to catch our attention.

It is a truth universally acknowledged that sentiment analysis is super fun, and Pride and Prejudice is probably my very favorite book in all of literature, so let's do some Jane Austen natural language processing.

Another day, another fuss about an animal study. This time, it's a cancer scare around a common antibacterial in soaps: triclosan. "The dirty side of soap," says the headline on the university's press release. "Triclosan, a common antimicrobial in personal hygiene products, causes liver fibrosis and cancer in mice." The title of the journal article it's promoting? The commonly used antimicrobial additive triclosan is a liver tumor promoter.

In a press release, the introductory paragraph is mostly made up of the answers to the W and H questions (see Section 2.1.1).

Next, use the following examples of strategies used in press-release introductions to adapt your introduction from Section 7.8.2 one more time to make it more appealing to a newsy audience. Now, instead of focusing on tone, you should focus on compressing a lot of information into a parsimonious, yet readable, couple of sentences.

A Catchy Title Plus an Abstract The title of a press release about the effect of a doctor's reassurance on recovery time begins with a title and no subtitle.

A doctor's reassurance speeds healing from an allergic reaction, find Stanford psychologists

The title is followed by a two-sentence abstract that stands in for a subtitle.

When doctors offer a few words of reassurance, patients suffering an allergic reaction begin to feel better more quickly. The finding by Stanford psychologists suggests that the placebo effect applies to words as well as to pills.

The first sentence of the abstract contains much of the same information that is in the title, but with more detail. The second sentence places the findings in a larger context of a placebo effect, which suggests why a limited study about allergic reactions could have greater importance.

7.8.10 Conclusions for Broader Publics

When writing more informally for a broader audience, the conclusion summarizes the take-away message. However, readers may not read through until the conclusion so do not rely on the end to introduce new material (recall the inverted pyramid of Section 2.1.3). If you think the information is critical for the reader to know, then move it to an earlier portion of the piece.

Consider the following examples of strategies used in blog post conclusions. Adapt your conclusion from Section 7.8.2 to make it more appealing to a broader audience.

Blog Post Conclusions That Remind and Look Forward The conclusion below reiterates the results and provides ideas for another data-collection procedure.

By collecting information on the people who I had lunch/dinner with and their suggestions for the place to eat for a whole week, I found that I had different habits for lunch and dinner. We were very likely to go to a place if we both suggested it. When we had more people together, it was easy to have disagreements. If I do this collecting procedure again, I would like to also collect the time of having lunch and dinner, and this may tell me whether the numbers of people and suggestions are related to the time.

The following conclusion gives a recommendation based on the analysis and reminds the reader of the motivation for studying indoor geolocation.

By deploying an indoor geolocation system in your store, not only can you locate goods in stock, but you can also help your customers shop at your store with high satisfaction. Moreover, with the inventory management software, you can also control your inventory efficiently. Finally, don't forget to place your Wi-Fi routers at the center of your store!

The next conclusion lists the things that the author learned while working on the data discussed in the post.

Here are some things I have learned while working with data on a relatively small organization:

1. Data quality is important, thus data documentation is necessary. In my case, we need to let future staff know why many sessions were starting only at the beginning of the minute. We should also specify that some users don't log into a desktop but still occupy desktop stations.
2. It is nice if you have good engineering skills, otherwise you have to rely on communication with data engineers to work with the data pipelines. In my case, I have to talk to the site managers or more senior staff to learn how the data is collected and wait for them to correct any inconsistencies.
3. Digital data is often not enough; you need to gather as much supplemental information as possible. In my case, besides our database data, I have lots of knowledge about the lab since I am physically in the lab for a long time.

In a press release especially, it can be useful to conclude with the larger context. Quotes are also used to conclude a press release since a researcher may be the best person to succinctly summarize the impact of the findings. Use the following examples to adapt your conclusion from Section 7.8.2 for a more news-oriented audience.

Press-Release Conclusions That Connect and Add Human Interest The conclusion for the press release about religious attendance and wellness explains how the featured finding connects to other work.

While previous studies of adult populations have found religious service attendance to have a greater association with better health and well-being than prayer or meditation, the current study of adolescents found communal and private spiritual practices to be of roughly similar benefit.

The press release about correctional officers and depression ends with a quote that explains future work inspired by their findings.

"Corrections is extremely difficult and emotionally demanding work," says Amy Lerman, the lead author of the study and a professor of public policy and political science. "We are just beginning to understand the huge range of mental and physical health issues that can result from exposure to violence and untreated toxic stress in the workplace. Agencies around the country are starting to look for ways to better support personnel—for the good of their employees and their families, the incarcerated population, and the system as a whole."

7.8.11 Drafting a Press Release

Press releases are written for journalists so that they can translate our work to the general public. The eventual audience consists of news readers who are broadly interested in science (including social, health, and physical sciences).

Common advice about writing for a broad audience is to write to an audience that is one level lower in education than the actual audience. Writing to a less-educated audience than our intended audience helps us limit the technical jargon to an appropriate level. For readers of science news stories, we take this advice to mean that we should write for a high-school science class, because the typical science news reader has a college education.

Use your storyboard from Section 6.7.5 to draft a press release. First, follow the steps below to extract relevant information for your press release.

- Go through your storyboard noting what is particularly interesting. What is the hook?

- Identify the five Ws. Who conducted the analysis? What were their findings from the analysis? Why are the findings important? When was the analysis carried out? Where was the analysis carried out?

- As appropriate identify who, where, and when for the study subjects.

- Identify the who, where, and when of the original data source, if the investigator didn't collect the data.

- Identify possible quotes. In a real press release, the authors would provide their own quotes and journalists would contact other sources for quotes. Here, instead, look for quotes in the supplementary materials given in Section 6.7.5 or find quotes in your own reliable sources.

- What additional details provide relevant information that support the story? Include details about the how of the investigation. Order these details from most to least important.

Now use the information that you collected to create a storyboard and draft a brief a press release. Organize your press release into the following sections/paragraphs.

- For the introduction, summarize the key information into a few sentences that include the Ws (refer to Section 7.8.9).

- Describe one or two of the most relevant details.

- Introduce a quote in support of these details or to introduce new details.

- Include additional relevant information that expands on the quotation (if it introduced these new details).

- Add a second quote that expands on any one of the themes already mentioned.

- Write a conclusion that summarizes what has been said already (refer to Section 7.8.10). End on a positive note that looks forward to the future.

7.8.12 Focus and Prior Knowledge for Different Audiences

For any piece of writing, there is usually some common foundation of knowledge that you expect your reader to have. A formal paper or report is typically aimed at professionals in your sub-field, and you assume the highest standard of prior knowledge. As you aim to widen your audience, you must expect less prior knowledge from your reader and account for this in how you explain or describe things. For example, unlike a formal report or a press release, the blogger gets to choose their audience. Blog posts can both be a place to talk about your work at a higher level or talk about the details that did not make it into the final report. The freedom to choose an audience brings with it the danger of failing to carefully identify the audience and their interests. To help choose a target audience consider what your goal is for writing the blog and remember that the more technical your writing is, the smaller the potential audience.

Consider the topic of a formal report you have written. For each of the following target audiences, write a few sentences describing the main focus of your writing if they were your audience.

- Professionals in your field but outside of your sub-field.

- Other professionals in a quantitative field.

- Those who have taken an introductory statistics course.

- General public (e.g. via a blog post).

For each audience type, list the related terms, techniques, or concepts that you assume they already know.

Materials for Section 7.8.2

Flight Problems	
Delay >0	41%
Cancelled	2%
Diverted	<1%

	Summary Statistics Per Airline Carrier		
carrier	count	mean delay (min)	mean distance
Hawaiian Airlines Inc.	366	20.78	2398.00
American Airlines Inc.	12008	17.75	1682.95
JetBlue Airways	1923	15.70	2027.34
Alaska Airlines Inc.	5062	14.97	607.10
Skywest Airlines Inc.	39027	14.76	306.93
United Air Lines Inc.	43736	14.07	1285.31
Southwest Airlines Co.	12568	13.36	568.40
Continental Air Lines Inc.	4801	13.16	2081.03
AirTran Airways Corporation	951	12.56	2090.73
Expressjet Airlines Inc.	1203	9.00	337.00
Frontier Airlines Inc.	1686	8.26	967.00
Delta Air Lines Inc.	4575	6.62	2148.50
US Airways Inc.	7112	5.93	1404.16
American Eagle Airlines Inc.	1742	4.94	372.00
Northwest Airlines Inc.	3827	4.10	1836.85

	Average Delay (min)	Average Delay (min) [within delayed flights]
Late Aircraft	29	32
Carrier	20	22
National Aviation System	10	9
Weather	1	1
Security	<1	<1

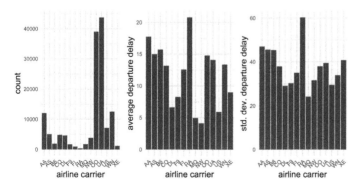

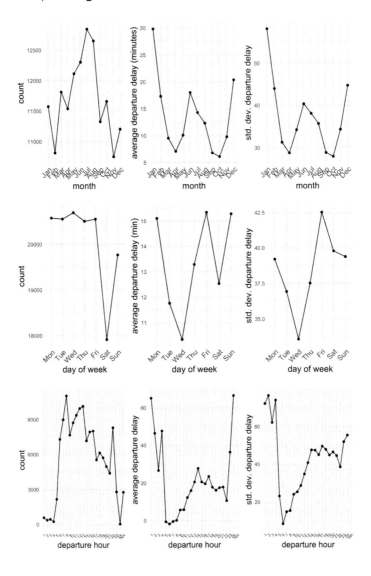

Seasonality in Flights by Top 4 Carriers in SFO

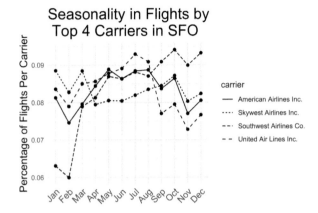

carrier
— American Airlines Inc.
···· Skywest Airlines Inc.
–·– Southwest Airlines Co.
– – United Air Lines Inc.

Overall Delay by Reason

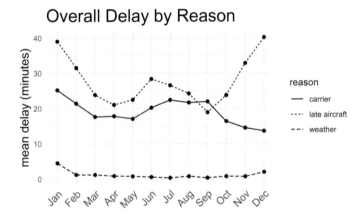

reason
— carrier
···· late aircraft
–·– weather

Within Delayed Flights: Delays by Reason

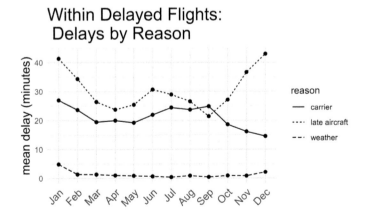

reason
— carrier
···· late aircraft
–·– weather

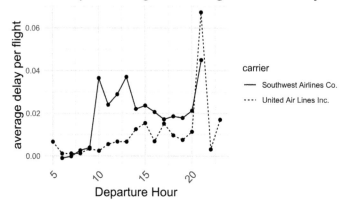

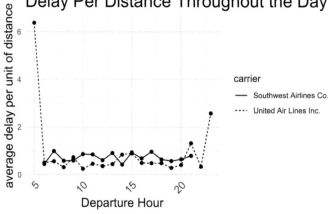

Part IV
Editing and Revising

Much of the time we spend writing is actually spent editing and revising. After we have written the first draft, we continue writing to fine-tune our arguments. We edit words and sentences, and we make larger changes where we reorganize, cut, and add new paragraphs and sections to our document.

When we write about a data analysis, we use statistical terms that can be particularly challenging to use properly. Part of the challenge is that often a statistical term has a meaning in everyday language which is not as precise as its statistical counterpart. Chapter 8 presents examples of both the misuse and proper use of statistical terms with the aim of helping you develop skills in statistical writing. Chapter 9 provides more general advice on how to edit your writing to reduce redundancy, add precision and clarity, and avoid grammatical pitfalls.

Chapter 10 examines the larger goal of revising an entire manuscript. We tackle challenges that might require reorganizing material, editing the document with a focus on a particular issue, and recognizing and filling holes in an argument. A variety of strategies for whole-document revisions are described there.

8 Taking Care with Statistical Terms

Data scientists face particular challenges in crafting clear sentences, choosing appropriate words, and conveying findings in a compelling manner that is faithful to the data and avoids overstating implications. This chapter offers advice on how to differentiate statistical terminology from everyday language, represent numbers in text, write mathematical expressions, and choose the correct quantitative nouns and adjectives (e.g., *fewer* or *less*, *percent* or *percentage*). Chapter 10 provides strategies for assessing how well you followed this advice in a first draft.

8.1 Statistical Terms and Everyday Usage

Many statistical terms give unique and specific meaning to words from everyday language. Examples of these include *confidence*, *error*, *sensitivity*, and *significant*. To reduce confusion, we recommend that you avoid these words in their common usage and exercise care when writing them in a technical setting, and depending on the audience, clarify the meaning of a term. In this section, we list several of the most common of these statistical terms and give examples of their proper usage. Section 9.2.3 gives general guidance for matching a word's connotation with the context.

Error An *error* in plain English typically describes a mistake or something wrong. *Statistical error* is quite different. It refers to a difference, such as the difference between an individual measurement and the average of several measurements. The terms *margin of error*, *measurement error*, *sampling error*, and *standard error* are a few examples of statistical error, each of which has a precise definition. Given so many related, but distinct, terms, it is important to clearly distinguish between them in your writing. For example, *The New York Times*' "Polling Standards" states that articles containing polling results should give the margin of error and "to aid comprehension

Communicating with Data: The Art of Writing for Data Science. Deborah Nolan and Sara Stoudt, Oxford University Press (2021). © Deborah Nolan and Sara Stoudt.
DOI: 10.1093/oso/9780198862741.003.0008

it [the margin of error] should be explained in a sentence." The example below shows what an explanation might look like.

Example: *Errors in a Smartphone Survey*
In the following report by the Pew Research Center on smartphone usage, the margin of error is provided for their survey. No explanation is provided because the audience is expected to be familiar with the concept.

Basic findings on smartphone ownership and demographics within the U.S. adult population are based on telephone surveys conducted December 4–7 and 18–21, 2014 among 2,002 adults. These figures for overall smartphone ownership/usage were collected via telephone survey to make them consistent with past Pew Research Center measurements of smartphone adoption. The margin of error for the total sample of 2,002 adults is plus or minus 2.5 percentage points. For more details about this survey, see Appendix A.

Another publication aimed at a less technical audience might include an additional sentence that briefly explains margin of error.

This means that in 95 cases out of 100, overall results based on such samples will differ by no more than 2.5 percentage points in either direction from what would have been obtained by seeking out all American adults. ■

Variance *Variation* in common usage typically refers to change or differences between things. The term *variance* has a specific statistical definition: an average (or expectation) of squared errors. Additionally, the term *variation*, while used more loosely than *variance*, also has a statistical connotation. *Variation* can refer to standard deviation, standard error, coefficient of variation, or variance, and it can refer to the less specific, yet still statistical, notions of variation that include bias, measurement error, and sampling error. Your writing should provide enough context to make it clear how you are using the word *variation*.

Example: *Coefficient of Variation*
Below, the coefficient of variation for two groups are compared.

For the iron retention data, the coefficient of variations are 0.69 and 0.67 for two types of iron. These data are quite "noisy" because the standard deviation is nearly 70% of the mean for both groups.

The coefficient of variation is defined implicitly in the second sentence; that is, we can figure out that it is the standard deviation as a proportion of the mean. ■

Confidence Of course, we want to be confident of our results, but in statistics *confidence* has a special meaning that pertains to interval estimates. Whether based on normal theory or bootstrap sampling, the general idea of a confidence interval is subtle, and the lay reader can easily miss the finer points. We advise staying close to well-established scripts when describing a confidence interval to avoid these pitfalls.

Example: *95% Confidence Interval from a Poll*
An example from a *New York Times* article in 2014 explains the confidence interval and coverage in terms of replicated polls.

This interval was calculated using a 95 percent confidence level, which theoretically means that if we conducted this same poll 1,000 times we would expect the interval to cover the true level of Mr. Cantor's support 950 times, more or less. ∎

Significance   There is a movement in the statistics community to reduce the use of the phrase *statistically significant*. This is in part because the word *significance* can mislead the reader into thinking the findings must be important, meaningful, or large. On the contrary, statistical significance need not imply any of these properties, for it is simply a descriptor of the size of a p-value.

The American Statistical Association (ASA) has published a statement cautioning the use of significance levels in testing and p-values. Their advice about p-values is helpful for writing about a data analysis. We provide four of ASA's points and comment on how they might apply to your writing.

- *P-values do not measure the probability that the studied hypothesis is true, or the probability that the data were produced by random chance alone.*
  Describe a p-value as the chance of observing data as extreme as yours under the assumptions of the studied hypothesis, i.e., under the assumptions of the probability model for generating the data.

- *Scientific conclusions and business or policy decisions should not be based only on whether a p-value passes a specific threshold.*
  Report the actual p-value, not simply whether or not the p-value exceeds a pre-specified level. Additionally, provide sample sizes and standard errors, for a more complete assessment.

- *Proper inference requires full reporting and transparency.*
  Discuss the full complement of data analyses carried out; mention models and tests that were attempted and abandoned because they did not provide significant results. In 100 hypothesis tests we expect 5 to be statistically significant when all of the null hypotheses are correct. A report of only those tests that produced small p-values gives a partial, inaccurate, and misleading story.

- *A p-value, or statistical significance, does not measure the size of an effect or the importance of a result.*
  Assess the importance of the effect found and supply additional information for the reader to make their own assessment. This extra information might include the size of an effect, sample size, standard deviation of the measurements, and possible non-probabilistic sources of error.

Example: *Significant Difference in Math Scores*

Suppose in a comparison of math scores between boys and girls, it was found that the average scores differed by a quarter of a point on a 100-point test, and the standard deviation of test scores was around 10 points for boys and girls alike. On a 100-point test with an SD of 10 points for individuals, a quarter-point difference in averages is tiny and neither important nor meaningful. However, if the numbers of boys and girls given the test are large, then the standard error of the difference will be very small and lead to a statistically significant difference between the two means. ■

Specificity, Sensitivity, and Precision These term are closely related, and they can be easily confused with each other and with their everyday counterparts (e.g., *sensitivity* often means reactivity in common usage). *Sensitivity* refers to the ability of a test to correctly reject a null hypothesis. *Precision* in, e.g., drug testing, addresses the question of the likelihood that someone who tested positive has the disease and so depends on the prevalence of the disease in the population. On the other hand, *specificity* describes the ability of a hypothesis test to correctly accept the null hypothesis. The example below demonstrates how a reminder of the definition of concepts can be interwoven in the description of findings.

Example: *Sensitivity and Precision of a Test for HIV*

The following description of a new test is argued to be better than the commonly used test because it can be easily and rapidly administered twice and so the repeated testing procedure has high precision.

A rapid test for HIV was developed for its sensitivity: the new test was shown to have sensitivity of 83%, compared to 60% for a commonly used test. This increased detection rate matters because those with false negative results (HIV-positive individuals whose disease goes undetected by a test) are likely not to return for further testing. The specificity of the rapid test is 97%. For example in a population of 692,000 with 12,000 HIV-positive, we expect 9,160 HIV-positive people to test positive and 20,400 HIV-negative people to test positive. That is, the test has a precision of only about 30%. However the rapid test can be immediately administered a second time for anyone with a positive result. With this two-step approach the precision increases to 94%. ■

Independent In common usage, *independence* often refers to freedom or individualism. In contrast, a statistical analysis typically focuses on describing the relationship between one *independent* variable and one or more dependent variables. Different fields have adopted conventions for describing this relationship. In machine learning, it is common to refer to the target and features or to the predicted and predictors. Other pairs of terms are the *explained* and *explanatory* variables, *regressand* and *regressors*, and *response* (or *outcome*) and *covariates*. Unfortunately, many of these pairs connote a causal relationship. The notion of explaining or predicting is not necessarily meant to be causal. Particularly confusing is the independent-dependent usage, and we recommend against using this pair of descriptors. For more details on causal connotations, see Section 8.5.

Confound One of the everyday usages of *confound* is close in meaning to its statistical counterpart: to mistake one thing for another. However, the statistical usage concerns making such a mistake in a very particular setting: when studying the relationship between a response and potential explanatory variables. Here a *confounder* can partly stand in for another explanatory variable and make it difficult to tease apart the relationship between the outcome of interest and other factors.

Example: Linking Cancer and the Pill
In a study of links between cancer and the contraceptive pill, potential confounders are carefully described.

When interpreting this estimate, however, it is important to remember that it is based on the crucial assumption that all of the difference was due to the effects of oral contraception. Although the RCGP study adjusted for potentially important confounding factors, i.e. age, smoking, social class, number of pregnancies and (when analysing the GP observation dataset) use of hormone replacement therapy, residual confounding other lifestyle or familial factors may be alternative explanations for at least part of the difference between groups.

Earlier in their report, the authors make the point that confounding is an ever-present problem for cohort studies, and observational studies more generally.

Such confounding can affect all observational studies. Although researchers can allow for potential confounding when analysing the data such adjustments can only be made for factors (confounders) that are measured. Sometimes the crucial confounders cannot be measured adequately or even at all. So, even after multiple adjustments have been made, residual confounding may account for some, or all, of an observed relationship between exposure and outcome. This makes it difficult to know whether a cause-and-effect link actually exists. ■

8.2 Help the Reader with Statistical Terms

Even though a data analysis is inherently technical, we want our writing to be understandable. Using technical terms and acronyms without first defining them alienates the reader. Examples and analogies ground the reader and can be an effective way to introduce new ideas.

Example: *Black-Box Terms*

The acronym *GLM* is used as a black-box term in the next sentence.

We use a GLM (generalized linear model) to predict whether an e-mail is spam or not.

Below, we rewrite this sentence to define the abbreviation and justify the use of the methodology.

Since the prediction of whether an e-mail is spam or not is a binary decision, we use a generalized linear model (GLM) that allows for a different distribution of errors than the traditional normal distribution. ∎

Example: *Generic Definitions*

Consider the following generic definition of statistical fraud detection that has little connection to the application at hand.

Statistical fraud detection methods may be supervised or unsupervised. Supervised methods use labeled data to learn patterns in each class. Unsupervised methods identify records that are far from the average in some way.

The alternative definition below includes examples and comparisons.

Statistical fraud detection methods may be supervised or unsupervised. In supervised methods, records known to be fraudulent or not are used to construct models to predict fraud. We use these models to assign new observations to one of the two classes. In contrast, unsupervised methods simply seek those accounts, customers and so forth which are most dissimilar from the norm. These unusual observations can then be examined more closely.

The connections to fraud detection help ground the reader. ∎

8.3 Similar Words with Distinct Meanings

When we write about data, we discuss percentages, mean effects, above and below average, approximate values, etc. Many times we are confronted with a choice between similar words, such as *percent* or *percentage*, *affect* or *effect*, *above* or *over*, and *about* or *approximate*. The precise definitions for each of these terms matters, and to write well, you need to select the correct word. In this section, we present several common confusions to look out for, and for each, we give examples of proper usage. Section 9.2.4 provides additional guidance about swapping in synonyms.

Data and Datum The singular term *datum* rarely appears in current publications in data science. A datum is the singular of data, but the use of *data* as a mass noun (i.e., a plural noun that can't easily be counted) has become widely accepted. To bridge the gap, some writers refer to the grammatically correct singular *data point* and *dataset*. Whether you choose to use *data* as a mass noun or the plural of *datum*, be consistent in your usage.

Example: *The Data is/are*
Given the modern usage of *data* as a mass noun, the following sentence is considered grammatically correct.

The data was collected over a number of years.

The next two sentences use *data* in its traditional plural form.

The data were cleaned before further analysis.
Any data point with a missing value was dropped from the dataset.

This last sentence refers to singular quantities, a point and a set, and is acceptable in both the traditional and current use of *data*. ∎

Relatedly, in stratified sampling, the population is partitioned into non-overlapping groups, called *strata*. The singular *stratum* refers to one of the subgroups.

Percent and Percentage A *percent* refers to a specific number, such as 20 percent, and a *percentage* refers to an unspecified portion, such as a large percentage. That is, *percentage* is used without numbers. When we take the difference between two percentages, that difference is measured in *percentage points*, e.g., 20 percent is 3 percentage points more than 17 percent.

Example: *Stats as Percentages*
The following sentences properly use *percent*, *percentage*, and *percentage point.*

Twenty percent of the eligible voters did not vote; this is a large percentage.
A greater percentage of eighth-grade students met the state standards than third-graders.
About 40 percent of mothers smoked when pregnant in the '60s compared to 8 percent today; that's a decrease of 32 percentage points, or an 80 percent decrease.

Although incorrect, the following use of *percent* has become commonly accepted.

What percent of your time do you spend watching TV?

The more grammatically correct usage is *percentage of your time.* ∎

Around, About and Approximately These words are used to express inexact values. The least precise is *around*, and it is typically not used in scientific settings. In technical settings, we typically use the term *about* to refer to imprecise quantities that are naturally measured in whole units, and we reserve *approximately* for measurements that are given in fractions.

Example: *Imprecise Figures*
A count takes on integer values so we use *about* with inexact count values. On the other hand, since an average value can be measured in fractional units, we use *approximately* to describe an inexact average value.

The dataset contains about ten missing values.
The average sale price of houses in the San Francisco Bay Area is approximately $400,000. ∎

Center on and Center Around When we refer to the proximity of a statistical estimator, such as a mean or median, to an ideal parameter value or a known constant, we use the phrase *center on*. As noted above, we use *about* and *approximately* to refer to numerical values. We never use the term *center around*.

Example: *Estimates and Estimators*
We describe a property of the sample mean with *center on*, and we use *approximately* for a value computed from data.

The distribution of the sample mean is centered on the true population mean. The sample mean is approximately 70 cm. ∎

Above and Over We normally use *over* not *above* with numbers. However, when we refer to numeric values in relation to a statistic, such as the mean or median, we use *above*. Temperature is a special case; degrees can be above or over a particular value, unless the comparison is to zero when we only use *above*.

Example: *Summary Stats*
The following sentences provide a variety of examples for where *above* and *over* are appropriate.

A newborn baby that weighs over 6 pounds is considered normal birthweight.
Fewer than 20% of the men had heights above average.
The majority of the survey respondents earn over $50,000 annually.
Last year, there were fewer days with temperatures above 80 degrees. ∎

Below and Under Similar to the distinction between *above* and *over*, we typically use *under* with numbers; *below* with statistics such as the average and median; and either with temperatures. Again, zero is a special case, where we only use *below*.

Example: *More Summary Stats*
The sentences for *above* and *over* in the previous example have been recast as examples for *below* and *under*.

A baby that weighs under 6 pounds is considered low birthweight.
Fewer than 20% of the men had heights below average.
The majority of the survey respondents earn under $50,000 annually.
Last year, there were fewer days with temperatures below 80 degrees. ■

Fewer and Less The canonical sign at a checkout stand in a grocery store reads (or should read): *12 items or fewer.* We use *fewer* when comparing quantities that can be counted, such as items for purchase, degrees of happiness, and inches of rainfall. On the other hand, we use *less* to compare quantities that cannot be counted, such as happiness.

Example: *How Happy?*
The first sentence in the quote refers to the concept of happiness and so uses *less* when comparing happiness as a mass noun. The second sentence refers to degrees of happiness, and so uses the term *fewer.*

In a happiness survey, respondents living in rainy climates tend to be less happy than those living in sunny climates.
In a happiness survey, respondents living in rainy climates tend to score fewer degrees of happiness than those living in sunny climates. ■

Example: *How Much Rain?*
The first sentence describes rainfall and density of a snow pack with the term *less* since we are not counting or measuring quantities. Alternatively, the second sentence refers to inches of rainfall so we use *fewer* instead of *less.*

Years when there is less rainfall, there is a less dense snow pack.
Years when there are fewer inches of rainfall, there is a less dense snow pack. ■

Affect and Effect Typically, we use *affect* as a verb that influences a noun. Since *affect* has causal implications, we should be cautious when using it. *Effect* is a noun that refers to a result or change. In statistics, the term *effect size* describes the magnitude of an effect, such as the size of a coefficient in a regression, the correlation between two variables, and the difference in means between groups. When we describe an *effect* we do not necessarily attribute causation to it.

Example: *Magnitude of an Effect*
The first sentence pertains to the coefficient of income in a model, and the second considers the relationship between age and sleep deprivation.

We compare the effect size of income on health across age groups.
What age group is most affected by sleep deprivation?

212 | Taking Care with Statistical Terms

The second sentence uses *affect* rather than *effect* because it refers to age as an influence on sleep deprivation. ■

Continual and Continuous The word *continual* implies something happening over and over again while *continuous* refers to something being uninterrupted or measured to potentially arbitrary precision.

Example: *Precision versus Repetition*
The elephant seal makes regular corrections to the direction of her path.

For three months, the female elephant seal continually corrects her path daily to travel along a great circle in search of food.

The normal curve is a smooth function.

The normal density curve is continuous.

Height can be measured to the nearest centimeter, decimeter, etc.

Height is a continuous variable. ■

Principle and Principal The word *principal* acts as both a noun and an adjective. When used as an adjective, *principal* refers to the primary or most important thing. In contrast, the word *principle* refers to a law or property.

Example: *Important Theories*
Principal component analysis (PCA) is a formal statistical technique that aptly employs the word *principal*. The second sentence below uses *principal* in a more general setting.

Principal Component Analysis can be an effective dimension reduction technique. The principal statistics of manufacturing are total revenue, total expenses, total salaries and wages, and cost of materials and supplies.

The following examples refer to theories and properties and so use *principle*.

We apply the least squares principle to fit our model.
The principle of redundancy is used when system reliability is improved through redundant or backup components.
The parsimony principle backs up this naïve view of causality. ■

8.4 Absolutes

In statistics, we rarely use absolutes because we typically have not observed all of the data. We know that there could be an exception that we have not had the opportunity to record. Even when we have observed all values for a population, we still remain cautious in our claims because others may want to generalize our findings to populations not (yet) studied. Examples of words that convey certainty and therefore should be avoided include *best*, *always*, and *never*. If you find yourself using them, ask yourself: best of what or might there be an exception? Adding a qualifier can remedy such overstatements, e.g., *best of the predictors we compared*.

Example: *Statistically Significant Covariate*
Below, the author states that socioeconomic status *always* has a significant coefficient.

Socioeconomic status was always statistically significant at the 0.001 level.

This claim might not be true for all possible models. The following revision provides the necessary caveat.

Among the models examined in this paper, socioeconomic status was always statistically significant at the 0.001 level. ∎

Example: *Unemployment Rates*
In the following statement, the author justifies their choice of the unemployment rate as a measure of the economy by stating it is the *most* common indicator.

The most common indicator of a healthy economy is the unemployment rate.

It would be difficult to establish the truth of this statement. The following revision makes a more reasonable claim:

A common indicator of a healthy economy is the unemployment rate.

This toned-down version is more than adequate for the purpose of the argument. ∎

8.5 Causal Statements

Words that imply a causal relationship all too easily slip into our descriptions of findings. There are two issues here. First, we may be making a conceptual error by attributing causality when the evidence does not support it. If the data you analyze are found data (i.e., data collected for a different purpose than the current analysis) or from an observational study, then you need to be especially careful how you report relationships and avoid drawing stronger conclusions than warranted.

The second issue is a language error. This occurs when we reach for a word and don't realize what meaning our choice brings along; it is easy to make the mistake of choosing a verb that implies causation. To help avoid this sort of error, we suggest that you conduct a separate pass in editing dedicated solely to spotting causal verbs. Common culprits are *leads to, brings about, due to, attributed to,* and *arises from.* We don't advise that you always substitute these phrases with the same standard phrase, e.g., *is associated with.* That would lead to rather boring prose. It can be a challenge to vary our language and stay away from the causal pitfall.

Even when we scrub causal statements from our writing, we often suggest explanations for our findings that carry causal implications. We advise a limit to these expositions because they tend to overreach. When you do include them, frame the causal explanation as an opinion and back it up with similar findings from other sound scientific research.

The key here is to be mindful of this sort of mistake and think about why we might make it. Is it that we do not understand the relationship we have discovered in our analysis? Mindful analysis of our argument is vital to avoiding these errors. Or, is it that we don't have the right word at hand? To help you navigate language errors, we have provided several examples. Each reports a relationship found in an analysis of observational or found data, where the descriptions use inappropriate causal language. In addition to the example of a too-strong claim, we offer ideas on how to write a more circumspect statement. The examples also try to demonstrate variety in how we might write about findings. See Sections 1.1.1 and 6.1 for examples and ideas on how to make your argument in a persuasive, yet accurate way.

Example: *Smoking and Birthweight*
In an analysis of the Child Health and Development Studies data on all pregnancies from 1960 to 1967 in the Kaiser Foundation Health Plan in the San Francisco–East Bay, one researcher wrote:

Smoking leads to low birthweight.

The researcher observed a correlation between smoking during pregnancy and low birthweight, but the use of *leads to* implies smoking causes low birthweight. The mothers in the study were not assigned at random to smoke or not while pregnant so we must be cautious in our claims. One alternative describes only the relationship found in the data.

Mothers who smoked during pregnancy tended to have lower birthweight babies compared to mothers who never smoked. ∎

Example: *Race Time and Age*
In another observational study, an analyst examined race results for participants in the Cherry Blossom Ten Mile Run. They describe their findings below.

Being 10 years older slows run time by 10 minutes on average.

If runners were followed over time, in a longitudinal study, then a claim such as this might be reasonable. However, these data do not follow runners as they age. The data represent a snapshot of run times for a group of people who are of different ages, e.g., the 30- and 40-year-olds are different groups of people. A better way to describe the findings is to report on the differences between the groups of runners, e.g.,

Comparing 30- to 40- to 50-year-old runners, we found run times decreased by 10 minutes on average from one group to the next. ∎

Example: *Eighth-Grade Math Scores*

An analysis of math scores of eighth graders in the Early Childhood Longitudinal Study, Kindergarten to Eight Grade, conducted by the National Center for Education Statistics, found a negative correlation between watching television and performance on math tests. The researcher writes:

Watching more than 4 hours of TV a day brings about lower math scores.

This statement implies causation. One way to check whether a sentence such as this makes a causal statement is to flip the statement around, e.g.,

Scoring poorly on math tests brings about excessive TV watching.

If the statement seems ridiculous, as is the case here, then we have a good indicator that the original version implies causation. An alternative is to simply report the direction of the relationship found between the two variables:

Watching more than 4 hours of TV a week is negatively associated with math scores.

Alternatively, we can make a more general statement such as *time spent watching TV is negatively associated with math scores*. ∎

Example: *Spam Detection*

A data analyst examined a collection of hand-classified emails (spam or not) made available by Spam Assassin. The analyst reported:

Spam arises when there are too many capital letters in an email.

The use of *arises when* seems a bit clunky. The author is searching for a way to describe the association between the frequency of capital letters in an email and whether or not it is spam. A more straightforward approach is to say:

A large number of capital letters in email is an indicator of spam.

or

A large number of capital letters in email indicates spam.

Indication of spam does not mean the capital letters cause spam. Given the data source, the notion of *causing* spam makes no sense. ∎

8.5.1 Avoiding Causal Exaggerations in Press Releases

We write press releases to call attention to our work. We aim to put our findings in the strongest terms so it can be easy to make claims that are stronger than justified. Additionally, when we try to write in simple compelling prose, it is easy to fall into the trap of asserting causality when our findings do not support such a claim. We provide an example of a press release that overreaches and another that expresses findings in a more cautious tone.

Example: *Causal Implications of Depression*
The first sentence of the press release about the correctional-officer survey makes a causal claim.

California's correctional officers are regularly exposed to traumatic events that make them more likely to grapple with depression, PTSD and suicidal thoughts, ...

This sentence catches our attention, the hook being the mental health of correctional officers. However, the sentence claims that exposure to traumatic events *make them* more likely to suffer from depression. The study is a survey so it is difficult to assert causality. We intuitively believe this to be the case and presumably other studies support the link between traumatic events and depression. However, a more accurate description of the findings would point out the link without implying causation. ∎

Example: *Links Between Religious Upbringing and Health*
The first paragraph of a press release about a study of children with different religious upbringings and their health in adolescence begins with a statement about how spiritual practice may be a protective factor for health in adolescence and early adulthood. The choice of *may be* is not a strong statement, but the press release is careful to not overstate the findings. The findings are described as an association, not a causal effect.

Participating in spiritual practices during childhood and adolescence may be a protective factor for a range of health and well-being outcomes in early adulthood, according to a new study from Harvard T. H. Chan School of Public Health. Researchers found that people who attended weekly religious services or practiced daily prayer or meditation in their youth reported greater life satisfaction and positivity in their 20s—and were less likely to subsequently have depressive symptoms, smoke, use illicit drugs, or have a sexually transmitted infection—than people raised with less regular spiritual habits. ∎

8.6 Numbers

There are a few basic rules for when to spell out a number and when to use numerals. We adopt the convention of spelling out positive integers under 10 and using numerals for larger values. Other conventions are also in use, such as using numerals for all numbers. Whichever convention you adopt, be consistent. The convention also should apply to words such as *first* and *second*. That is, if your convention is to write *1* rather than *one*, then you should also write *1st* rather than *first*.

Decimal numbers always should be expressed with numerals, e.g., 3.14. If the number is between −1 and 1, then provide a leading zero before the decimal point, e.g., 0.67 and −0.15. The leading zero helps the reader notice the decimal point.

Lists of Numbers Even with a convention, there are exceptions to when numbers are spelled out. For example, a list of numbers should always be presented as numerals.

Example: *Some Typical Values*
At times, we present snippets of data to help the reader understand the typical values.

The daily rainfall measurements for one week were 1, 12, 14, 0, 0, 0, 0 hundredths of an inch.

We do not spell out the numeral 1 in the above sentence because it appears in a list with other numbers. ∎

Proximity We also break with convention when two numbers appear next to each other in a sentence because it is easier to read if we spell one of them out and use numerals for the other.

Example: *Numbers of Variables and Models*
The sentence below contains both 15 and 3. (Note that we don't spell out the numeral 3 in the previous sentence because it appears with the numeral 15 in a list of sorts.) Our convention tells us to spell out *three* and use numerals for 15.

We fitted 15 three-variable models.

Furthermore, although seven and three are both less than 10, we spell one out for readability.

We fitted seven 3-variable models.

Note that *fit* or *fitted* are both commonly used as the past tense of the verb to fit. Choose one and stick to it. ∎

Starting a Sentence When a number appears at the beginning of a sentence, we spell out the number no matter its size. If you don't want to spell it out because, e.g., it's so large, then rewrite your sentence so that the number appears later in the sentence.

Significant Figures Often statistical software presents results from a calculation with many more digits than we need. When we include summary statistics from these calculations in our reports, we should report only the relevant digits. The standard deviation of the data gives a good indication of the digits to drop. See Section 3.3 for an example.

8.7 Mathematical Expressions

When we write mathematical symbols and expressions, we follow the standard grammatical rules for written English. That is, mathematical symbols correspond to different parts of speech and collections of them form clauses and sentences. For example, the equation, $x + y = 10$, is read as x *plus* y *equals ten*. Here, we use the symbol $=$ as a verb. Mathematical clauses and sentences need punctuation such as commas, semicolons, and periods to be grammatically correct and to help the reader understand the mathematics.

As an example, consider the following sentence:

When $x < 0$, then $f(x)$ is strictly increasing.

Here, the expression $f(x)$ is treated as a noun, and *When* $x < 0$ is a clause. We read the entire sentence as

When x is less than 0, then f of x is strictly increasing.

In addition to punctuation, the words surrounding the symbols are important because they help readers understand the mathematics.

There are a few basic rules for writing mathematics that help with readability.

- Do not begin a sentence with a mathematical symbol. Although the sentence may be grammatically correct, it can be jarring to the reader and hard for them to understand.

- Define a mathematical symbol before (or at the same time as) using it in an expression.

- Avoid using letters for mathematical symbols that can be confused with English words. For example, the symbol a is a poor choice for a constant.

- Avoid using mathematical symbols in titles and headings.

- Some aspects of mathematics are easier to understand when written out, while others are more easily comprehended in mathematical notation. Choose the representation accordingly.

- Once you use a word to represent a mathematical concept, don't use a synonym as an alternative for the term. The word is temporarily associated with a precise mathematical meaning and substituting a similar work for this term is confusing.

We recommend that you edit your mathematics writing by reading it out loud. It can be easier to hear grammatical errors than to recognize them in print. If you hear yourself saying a series of fragmented sentences, then you need to edit your writing.

Example: *Model for Donkey Weight*
Below is a description of a general model for the weight of a donkey.

Suppose that three quantities x, y, and z are related in the form:

$$f(x)+g(y)=h(z).$$

Here x and y could be the height and girth of the donkey, and z could be its weight. The functions f, g, and h are monotonic (i.e., always increasing or always decreasing).

The display of the model on its own line, rather than embedded in a line of text, signals the reader that the equation is important. The period appearing after $h(z)$ in the displayed equation denotes the end of the sentence, and the English description of x, y and z as *three quantities* makes the sentence a bit easier to read. Also, *Here* at the beginning of the second sentence and *The functions* at the start of the third keep these sentences from beginning with mathematical symbols. This helps with the flow. Below we remove these helper words for comparison. Although this second version is grammatically correct, the first is easier to read.

Suppose x, y and z are related in the form:

$$f(x)+g(y)=h(z).$$

x and y could be the height and girth of the donkey, and z could be its weight. f, g, and h are monotonic (i.e., always increasing or always decreasing). ∎

Example: *Meta-Analysis Model*
Consider the following description of the model discussed in Possolo et al. (see Section 1.2.2 for more details).

The statistical model underlying the DerSimonian-Laird procedure is an additive, random effects model, which represents each measured value as a sum of three elements, $h_j = h + \lambda_j + \epsilon_j$, for j = 1, ..., n different experiments, where h denotes the measurand, the $\{\lambda_j\}$ denote experiment (or laboratory) effects, and the $\{\epsilon_j\}$

represent experiment-specific measurement errors. ...The $\{\lambda_j\}$ are modeled as values of independent random variables with mean 0 and standard deviation τ > 0.

Notice that the model is not displayed on its own line, but appears in-line in the paragraph. The authors may have considered the model to be simple enough to not require calling attention to it in a displayed equation. The phrase *a sum of three elements* is technically unnecessary but helpful. Notice that the three elements are defined in the same sentence where they are first used. Also notice that rather than use the mathematical expressions $\mathbb{E}(\lambda_j) = 0$ and $\mathrm{SD}(\lambda_j) = \tau$, the authors use words to provide the expected value and standard deviation of the quantity λ_j. This use of plain English is a common approach and is easier to read. Additionally, when using mathematical symbols, we cannot write $\mathrm{SD}(\lambda_j) = \tau > 0$ because it is not grammatically correct; that is, *SD of lambda sub j equals tau is greater than 0* doesn't make sense. ∎

With mathematics, it is good practice to include *then* in a statement such as the following.

If x > 30, then 0 is the minimizer.

The sentence below is identical except for the appearance of *then*.

If x > 30, 0 is the minimizer.

While these sentences are identical in meaning, the reader has to work harder to figure out the connection between the numbers 30 and 0 in the second version.

8.7.1 Formatting Mathematics

Journals have guidelines for formatting mathematics including advice about when to display and number an expression, when to use exponents, and how to write fractions. In this section, we provide a few basic rules that are common in practice.

There are three characteristics to consider when deciding whether or not to display a mathematical expression on its own line: its importance, its length, and its height. Important expressions should be displayed separately from the text so that they cue the reader as to their importance and easily stand out for the reader to back-reference. A long expression should also be displayed to avoid breaking it across a line of text, which can make it difficult to read. Likewise, tall expressions should be displayed. When a fraction is in-line, then we write it as, e.g., $(x + y)/(y + z)$, and when displayed, we express it in its tall form as

$$\frac{x + y}{y + z}.$$

Superscripts and subscripts also add to the height of an expression. Tall expressions that have an extended exponent should be displayed. Double subscripts should be avoided. In the special case of e, it should be written as $\exp()$ if the exponential is long, whether or not it is displayed.

Avoid numbering a displayed equation unless it is important and needs to be referred to later. When multiple lines of mathematics are displayed, they should be aligned. That is, the connective, e.g., $=$ or $\leq$, should align from one line to the next. Also, if a single equation needs to be displayed on more than one line then it should break at say a "+" and indented slightly from the "=". Below is an example,

$$x = a + b + c$$
$$+ d + e$$
$$< f$$

Notice that the first equation is split over two lines with the second portion indented past the "=", and the inequality is aligned with the first equation.

As mentioned earlier, constants and variables should be defined when they are first used. When there are several to be introduced, consider placing the variables and their definitions in a table for easy reference.

Statisticians tend to consistently use specific letters of the alphabet for constants, functions, indices, and random variables. That is, the letters a through d are typically reserved for constants and parameters; f, g, and h for functions; i, j, and k for indices; m and n for sample sizes; p and q for probabilities (p is also used for the number of parameters or dimension); and x, y, and z for observed data. We also use the capital letters X, Y, and Z for random variables. Some letters in the Greek alphabet also tend to have specific uses, e.g., μ for expected value, σ for standard deviation, and θ for a generic parameter. Other fields have different conventions. For example, in econometrics, t, used for time, is a typical index, and in computer science d is used for dimension. It's a good idea to adopt the conventions of your audience when choosing notation.

Mathematical symbols use a different font style, usually italics, to distinguish the mathematics from plain English. This change in font happens automatically when the mathematics are properly marked in a document preparation system such as LaTeX and Markdown.

The guidelines we have provided in Sections 8.6 and 8.7 may seem overwhelming. It appears that there are many rules to remember. We suggest that rather than memorize these conventions, you just remember that when you write, you have a reader. Be kind to your reader; adopt formatting for numbers and mathematical expression that lower the cognitive load for them.

8.8 Notes

The New York Times' "Polling Standards" for describing a margin of error can be found at [25]. The Pew Research Center report on smartphone usage was made by Smith [21]; the coefficient-of-variation example comes from Rice [17]; and the confidence-interval example comes from Vavreck [30].

The American Statistical Association's statement on the use of significance levels in testing and *p*-values can be found at [2].

The example of sensitivity and specificity for HIV tests is based on the ideas in [7]. Consult Hannaford [10] for the discussion of the study of links between cancer and the contraceptive pill described in the example of confounding.

The example of generic definitions was adapted from text in Bolton and Hand [6].

Many of the terms in Section 8.3 were adapted from material in Alley's *The Craft of Scientific Writing* [1].

The examples in Sections 8.4 and 8.5 were inspired by student work. The press releases referred to in Section 8.5.1 came from the press offices of UC Berkeley [28] and Harvard [24] respectively.

For advice on numbers and mathematical expressions, we consulted and adapted material from Lee's "A guide to writing mathematics" [12], Maurer's "Short guide to writing mathematics" [14], and Gelman's blog post [9].

The press releases used in Section 8.10.5 are by Anwar [3] and Sanders [18], respectively. The press release in Section 8.10.6 is by Anwar [3], which is based on the article by Smarr and Schirmer [20]. The blog used as an example in Section 8.10.8 is by Barter [4].

For more practice, consider working on Portfolio Pieces J and S.

8.9 References

[1] M. Alley. *The Craft of Scientific Writing*. Springer-Verlag, New York, 3rd edition, 1996.

[2] American Statistical Association. American Statistical Association releases statement on statistical significance and p-values. https://www.amstat.org/asa/files/pdfs/P-ValueStatement.pdf, March 2016.

[3] Y. Anwar. Stressed to the max? Deep sleep can rewire the anxious brain. Media Relations, UC Berkeley, November 2019. https://news.berkeley.edu/2019/11/04/deep-sleep-can-rewire-the-anxious-brain/.

[4] R. Barter. ANOVA. https://rlbarter.github.io/Practical-Statistics/2017/02/20/anova/, February 2017.

[5] BBC. The challenge of "farming the desert" in Australia. https://www.bbc.com/news/world-australia-45179181, September 2018.

[6] R. J. Bolton and D. J. Hand. Statistical fraud detection: A review. *Statistical Science*, 17(3):235–249, 2002.

[7] Centers for Disease Control. Notice to readers: Approval of a new rapid test for HIV antibody. *Morbidity and Mortality Weekly Report*, 51(46):1051–1052, 2002. https://www.cdc.gov/mmwr/preview/mmwrhtml/mm5146a5.htm.

[8] K. E. Cortes and H. Fricke. Text messages to parents can help boost children's reading skills. *The Conversation*, August 2018. https://theconversation.com/text-messages-to-parents-can-help-boost-childrens-reading-skills-100306.

[9] A. Gelman. Advice on writing research articles. https://statmodeling.stat.columbia.edu/2009/07/30/advice_on_writi/, April 2018.

[10] P. Hannaford. Investigating the link between the pill and cancer. *Significance*, 5(1):6–10, 2008.

[11] Z. Kleinman. Do video games make people violent? BBC News, August 2015. https://www.bbc.com/news/technology-33960075.

[12] K. P. Lee. A guide to writing mathematics. https://web.cs.ucdavis.edu/~amenta/w10/writingman.pdf.

[13] L. Martinez, E. McLaughlin, and M. Keneally. North Korea says it tested new, nuclear-capable ICBM that can reach continental US. ABC News, November 2017. https://abcnews.go.com/International/north-korea-fires-1st-missile-months-south-korean/story?id=51437178.

[14] S. B. Maurer. A short guide to writing mathematics. http://www.swarthmore.edu/NatSci/smaurer1/WriteGuide/.

[15] D. Montanaro. Here's just how little confidence Americans have in political institutions. NPR, January 2018. https://www.npr.org/2018/01/17/578422668/heres-just-how-little-confidence-americans-have-in-political-institutions.

[16] S. A. O'Brien. San Francisco pressures Uber and Lyft over how they pay drivers. CNN Business, May 2018. https://money.cnn.com/2018/05/29/technology/uber-lyft-driver-wages/index.html.

[17] J. Rice. Mathematical Statistics and Data Analysis. Cengage Learning, Boston, MA, 2013.

[18] R. Sanders. Early climate modelers got global warming right, new report finds. Media Relations, UC Berkeley, December 2019. https://news.berkeley.edu/2019/12/04/early-climate-modelers-got-global-warming-right-new-report-finds/.

[19] ScienceDaily. New statistical approach will help researchers better determine cause-effect. https://www.sciencedaily.com/releases/2016/06/160621155002.htm, September 2016.

[20] B. L. Smarr and A. E. Schirmer. 3.4 million real-world learning management system logins reveal the majority of students experience social jet lag correlated with decreased performance. Scientific Reports, 8(4793), 2018. https://doi.org/10.1038/s41598-018-23044-8.

[21] A. Smith. U.S. smartphone use in 2015. https://www.pewresearch.org/internet/2015/04/01/us-smartphone-use-in-2015/, April 2015.

[22] P. Smith and J. L. Zagorsky. Does thinking you look fat affect how much money you earn? The Conversation, July 2018. https://theconversation.com/does-thinking-you-look-fat-affect-how-much-money-you-earn-99262.

[23] S. Stoudt. Fixed, mixed, and random effects. https://rlbarter.github.io/Practical-Statistics/2017/03/03/fixed-mixed-and-random-effects/, March 2017.

[24] C. Sweeny. Religious upbringing linked to better health and well-being during early adulthood. Harvard T. H. Chan School of Public Health, September 2018. https://www.hsph.harvard.edu/news/press-releases/religious-upbringing-adult-health/.

[25] The New York Times. Polling standards. http://www.nytimes.com/packages/pdf/politics/pollingstandards.pdf, June 2006.

[26] J. M. Twenge. Have smartphones destroyed a generation? The Atlantic, September 2017. https://www.theatlantic.com/magazine/archive/2017/09/has-the-smartphone-destroyed-a-generation/534198/.

[27] Uber. Making our roads safer—for everyone. https://www.uber.com/blog/making-our-roads-safer-for-everyone-2/, January 2015.

[28] UCB Public Affairs. Correctional officers at high risk for depression, PTSD, suicide, survey finds. Media Relations, UC Berkeley, April 2018. http://news.berkeley.edu/2018/08/23/california-correctional-officers-at-high-risk-for-depression-ptsd-and-suicide-new-survey-finds/.

[29] US News & World Report. Errors found in plea deals of 3 convicted in man's slaying. https://www.usnews.com/news/best-states/north-carolina/articles/2018-02-15/errors-found-in-plea-deals-of-3-convicted-in-mans-slaying, February 2018.

[30] L. Vavreck. Why polls can sometimes get things so wrong. *The New York Times*, July 2014. https://www.nytimes.com/2014/07/04/upshot/why-polls-can-sometimes-get-things-so-wrong.html.

[31] Wikipedia. List of statistics articles. https://en.wikipedia.org/wiki/List_of_statistics_articles.

[32] Wikipedia. Analysis of variance. https://en.wikipedia.org/wiki/Analysis_of_variance, April 2020.

8.10 Activities

8.10.1 Finding Statistical Terms in Everyday Language

Read a few news articles on any topic and see if you can find some of the statistical terms discussed in Section 8.1 used in their everyday sense. What confusion could arise if the terms were interpreted in their statistical sense? If you need a place to start, check out the following.

"North Korea says it tested new, nuclear-capable ICBM that can reach continental US" (Martinez et al. [13]).

"Here's Just How Little Confidence Americans Have In Political Institutions" (Montanaro [15]).

"Errors Found in Plea Deals of 3 Convicted in Man's Slaying" (U.S. News & World Report [29]).

"The challenge of 'farming the desert' in Australia" (BBC [5]).

"San Francisco pressures Uber and Lyft over how they pay drivers" (O'Brien [16]).

8.10.2 Practice Distinguishing Between Similar Words

Pick at least three of the groups of similar words from Section 8.3 and write a sentence for each word that shows you understand the distinction between them.

Next, fill in the blanks with the appropriate word:

- By rewriting the algorithm, we can increase efficiency and use _____ computational resources. (few/less)

- We are aiming for an unbiased estimator to ensure that our estimate is _____ the truth. (centered around/centered on)

- For regions with mountains _____ 20,000 feet, our model works less well. (above/over)

- We looked further into the standardized residuals _____ two. (below/under)

- The difference in the approval _____ between the two candidates was less than one _____. (percent/percentage/percentage point)

- The study aimed to determine the _____ of smoking on deep sleep. (affect/effect)

- Our data is not _____ so we cannot use ordinary least squares. (continual/continuous)

- The typical height we saw for women was _____ 5′ 4″. (around/about/approximately)

- The increase in number of participants over time is the _____ pattern we find. (principle/principal)

8.10.3 Focused Revision at Sentence Level

We provide several sentences that need to be revised. For each sentence, decide which of the five issues in the bulleted list has occured and write a revised version that addresses the problem. Refer to Section 10.2 for more guidance on targeted revision. (There may be more than one problem with a sentence).

- *Absolutes and Causal Statements.* Look for overreaching conclusions.

- *Statistical Terms and Everyday Language.* When statistical terms are used, is it clear when they are being used statistically versus in the everyday sense?

- *Similar Words with Distinct Meanings.* Are any common mistakes made concerning words in Section 8.3?

- *Numbers.* Are numbers dealt with according to convention?

- *Help with Statistical Terms.* Look for black-box terms and generic definitions that could be made more precise or relatable.

1. The number of visits drops significantly after children turn 6 and increases gradually until they reach the age of 55. The number of visits has another significant drop after the age 55.

2. Suicide attempts are most likely to occur from 6 pm to midnight during a day, and most likely to occur from April to June during a year. Only 5 percent of ER visits are suicide attempts.

3. With the legalization of marijuana in certain regions, there has been a spike in consumption, especially among teenagers.

4. Further data visualizations of age show an increase in substance usage towards the teenage years and reveal a slight positive skew.

5. We found that the use of pharmaceuticals is constantly increasing along with age after 20, while the abuse of illicit drugs and opiates is consistent at a low percent among people of all ages.

8.10.4 Writing and Speaking Mathematics

A good way to improve our mathematical writing is by reading it aloud. Your ears can often pick out sentence fragments and grammatical errors better than your eyes. If you find yourself saying a series of fragmented sentences and equations, you should do some rewriting. Below is a paragraph with mathematical content. Read it out loud and identify expressions that should be displayed and sentence fragments. In addition to fixing these problems, add words to help with comprehension and include definitions of symbols. To help with this, the symbol σ represents a nonnegative-scale parameter and α a shape parameter. You can assume that the random variable Z has been defined earlier.

In this case, the tail of the distribution is characterized by the equation $P(Z > z + u) = (1 + \alpha \frac{z}{\sigma_u})_+$. $a_+ = a$, $a \geq 0$ and $a_+ = 0$, $a < 0$. α controls the tails of the distribution.

8.10.5 Statistical Terms in Press Releases

Rarely do we need to describe methods in a press release with advanced statistical terminology. The press release should keep the method descriptions at a high level, unless the method is an essential part of the findings. Even then, the methods should be described in general terms that use examples and analogies when possible. Some news outlets provide guidelines for how to use common statistical terms such as *confidence interval* and *p-value*. See Section 8.1 for an example. Furthermore, Section 8.1 explains how we need to be careful to differentiate statistical terms from common, everyday language, and this distinction is vital in a press release where we are expected to write using plain English.

Below we provide a sentence from two press releases that contain statistical terms. Whether or not the term is used in the statistical sense or the everyday sense is unclear. Rewrite each sentence two ways, once assuming the term is used in the statistical sense and once assuming the term is used in the everyday sense.

After a full night of sleep, during which participants' brain waves were measured via electrodes placed on their heads, the results showed their anxiety levels declined significantly, especially for those who experienced more slow-wave non-rapid eye movement sleep.

"The real message is that the warming we have experienced is pretty much exactly what climate models predicted it would be as much as 30 years ago," he said. "This really gives us more confidence that today's models are getting things largely right as well."

8.10.6 Causal Statements in Press Releases

Figure 8.1 contains a press release that reports on a study of college students' biological clocks and their grades.

The first five paragraphs in the press release have been numbered. For each paragraph determine whether a finding is described as a causal effect or an association (see Section 8.5 for assistance in identifying causal claims). For those instances that imply causality, rewrite the sentence to weaken the claim but maintain its strength as much as possible. Some of these paragraphs have descriptions of more than one finding. Be sure to consider all of them.

8.10.7 Looking for Causal Statements

Read a few news articles (science articles may be more fruitful) and identify any causal statements that are overreaching. Rewrite the sentence to better reflect the scope of the results. If you cannot find any statements that are overreaching, identify appropriate use of causal statements and explain why they are appropriate. If you need a place to start, try the following articles.

"Have smartphones destroyed a generation?" (Twenge [26]).
"New statistical approach will help researchers better determine cause-effect" (ScienceDaily [19]).
"Does thinking you look fat affect how much money you earn?" (Smith and Zagorsky [22]).
"Do video games make people violent?" (Kleinman [11]).
"Making our roads safer— for everyone" (Uber [27]).
"Text messages to parents can help boost children's reading skills" (Cortes and Fricke [8]).

8.10.8 Blogging to Teach a Statistical Concept

Blog posts provide a venue for someone with expertise to teach a concept to a less technical audience. Writing a blog post that walks through the basics can serve as an easy entry point for someone who may not have the time or resources to formally learn on their own. However, even if we are an expert on a statistical concept, we may uncover gaps in your understanding as we struggle to explain a concept simply.

In the following excerpt of a blog post, the author explains the essential facts about a commonly used statistical tool, analysis of variance (ANOVA). We can use the strategies for reading blogs learned in Section 2.3 to analyze the text. First, the writer motivates learning about ANOVA by unpacking the meaning of its name.

Poor grades tied to times that don't match our biological clocks

By Yasmin Anwar, Media Relations | MARCH 29, 2018

[1] It may be time to tailor students' class schedules to their natural biological rhythms, according to a new study from UC Berkeley and Northeastern Illinois University.

[2] Researchers tracked the personal daily online activity profiles of nearly 15,000 college students as they logged into campus servers. After sorting the students into "night owls," "daytime finches" and "morning larks"—based on their activities on days they were not in class—researchers compared their class times to their academic outcomes.

[3] Their findings, published today in the journal *Scientific Reports*, show that students whose circadian rhythms were out of sync with their class schedules—say, night owls taking early morning courses—received lower grades due to "social jet lag," a condition in which peak alertness times are at odds with work, school or other demands.

[4] "We found that the majority of students were being jet-lagged by their class times, which correlated very strongly with decreased academic performance," said study co-lead author Benjamin Smarr, a postdoctoral fellow who studies circadian rhythm disruptions in the lab of UC Berkeley psychology professor Lance Kriegsfeld.

[5] In addition to learning deficits, social jet lag has been tied to obesity and excessive alcohol and tobacco use. On a positive note: "Our research indicates that if a student can structure a consistent schedule in which class days resemble non-class days, they are more likely to achieve academic success," said study co-lead author Aaron Schirmer, an associate professor of biology at Northeastern Illinois University.

Not only night owls at risk

While students of all categories suffered from class-induced jet lag, the study found that night owls were especially vulnerable, many appearing so chronically jet-lagged that they were unable to perform optimally at any time of day.

But it's not as simple as students just staying up too late, Smarr said.

"Because owls are later and classes tend to be earlier, this mismatch hits owls the hardest, but we see larks and finches taking later classes and also suffering from the mismatch," said Smarr. "Different people really do have biologically diverse timing, so there isn't a one-time-fits-all solution for education."

In what is thought to be the largest-ever survey of social jet lag using real-world data, Smarr and Schirmer analyzed the online activity of 14,894 Northeastern Illinois University students as they logged in and out of the campus's learning management system over two years.

To separate the owls from the larks from the finches, and gain a more accurate alertness profile, the researchers tracked students' activity levels on days that they did not attend a class.

Majority out of sync with class times

Next, they looked at how larks, finches and owls had scheduled their classes during four semesters from 2014 to 2016 and found that about 40 percent were mostly biologically in sync with their class times. As a result, they performed better in class and enjoyed higher GPAs.

However, 50 percent of the students were taking classes before they were fully alert, and another 10 percent had already peaked by the time their classes started.

Previous studies have found that older people tend to be active earlier while young adults shift to a later sleep-wake cycle during puberty. Overall, men stay up later than women, and circadian rhythms shift with the seasons based on natural light.

Finding these patterns reflected in students' login data spurred researchers to investigate whether digital records might also reflect the biological rhythms underlying people's behavior.

The results suggest that "rather than admonish late students to go to bed earlier, in conflict with their biological rhythms, we should work to individualize education so that learning and classes are structured to take advantage of knowing what time of day a given student will be most capable of learning," Smarr said.

Figure 8.1 Press Release on the Relationship between Grades and Sleep.

The writer exercises empathy in this process by asking and answering a fundamental question that a reader might have but thought too naive to ask.

Why is ANOVA called "Analysis of Variance"?
What does comparing means have to do with variability?

Quite a lot it turns out ...simply by asking "are the means different", we are essentially asking a question about whether the variance of the means is large. However, the variability that we observe between the means themselves only makes sense relative to the overall variance in the data.

Once the motivation has been established and the role of variance has been introduced, it is important to distinguish between types of variance. This blog excerpt closes with a recap of two different types of variability, along with accessible definitions, and an easy-to-understand summary of how they are used in the reasoning for the ANOVA method. By focusing on the fundamental details rather than the technical details, the reader gets the big-picture idea.

There are two types of variance at play here:

• within-group variability: the variance of the individual observations within a group, and

• between-group variability: the variance between the averages of the groups.

...The basic idea is that if the variability between the groups is greater than the variability within the groups, then we have evidence that the differences between the groups is not simply reflecting random noise.

How does this blog explanation differ from how ANOVA was introduced to you? If you are not familiar with ANOVA, compare these blog excerpts to the Wikipedia page for ANOVA [32].

Search for another blog post that describes a statistical or scientific concept at a high level. If you need a place to start, try Stoudt's blog post on fixed, mixed, and random effects [23]. What properties of the post make it effective in explaining the concept to a broader audience? What could be improved to make the content more precise or accessible? Consider the choice of examples and level of formality used.

8.10.9 Edit a Statistical Wikipedia Entry

Writing or editing a statistics entry in Wikipedia provides valuable experience with writing technical material for a broad audience. There are many ways to edit an entry.

• Add an example. Would a diagram or other graphic help explain the topic? Is there a motivating example you could propose and work through? Make sure the setup of your example, how it connects to the topic, and its final take-away are clearly explained.

- Clarify content. When you read through the pre-existing content, where do you have to pause or get confused? Are any non-standard terms used that lack a link to another Wikipedia page or are not defined? Even small changes can help improve the clarity of the entry.

- Add more information to a stub. Focus on the introductory sections that are written for an audience unfamiliar to the topic. What are the key definitions, intuitive examples, and other background needed for a reader to understand the topic at least at a high level?

- Fact check. Content on Wikipedia is often technically correct. However, there is often room to make the content more precise. Look at the more technical sections of the entry. Is anything glossed over? Is a key detail omitted that someone with more familiarity with the topic would detect?

Choose an entry from the list of statistical Wikipedia entries [31]. Select a term you feel comfortable explaining to a peer. Make one or more of the above types of revisions. Make sure your additions are technically correct, free of grammar and spelling errors, relevant to the topic, and help clarify the topic for a reader with less experience in statistics than you.

Standard error

From Wikipedia, the free encyclopedia

> This is an old revision of this page, as edited by Derek R Bullamore (talk I contribs) at 19:31, 11 November 2017 *(Filled in 2 bare reference(s) with reFill ())*. The present address (URL) is a permanent link to this revision, which may differ significantly from the current revision.

(diff) ← Previous revision I Latest revision (diff) I Newer revision → (diff)

For the computer programming concept, see standard error stream.

The **standard error (SE)** of a statistic (most commonly the mean) is the standard deviation of its sampling distribution,[1] or sometimes an estimate of that standard deviation.

The equation for the **standard error of the mean (SEM)** depicts the relationship between the dispersion of individual observations around the population mean (the standard deviation), and the dispersion of sample means around the population mean (the

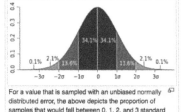

For a value that is sampled with an unbiased normally distributed error, the above depicts the proportion of samples that would fall between 0, 1, 2, and 3 standard deviations above and below the actual value.

standard error). Different samples drawn from that same population would in general have different values of the sample mean, so there is a distribution of sampled means (with its own mean and variance). The relationship with the standard deviation is such that, for a given sample size, the standard error equals the standard deviation divided by the square root of the sample size. As the sample size increases, the sample means cluster more closely around the population mean and the standard error decreases.

In regression analysis, the term "standard error" is also used in the phrase standard error of the regression to mean the ordinary least squares estimate of the standard deviation of the underlying errors.[2][3]

Figure 8.2 Screenshot of the Lead Section of the Wikipedia Article on Standard Error. The lead section immediately follows the title. The first sentence defines the standard error in terms of the standard deviation of the sampling distribution of a statistic. Readers who are unfamiliar with these terms can click through to cross-referenced articles.

To get you started, below are some guided edits for you to make to the Wikipedia page on *standard error*. A screenshot of the section appears in Figure 8.2.

- The first sentence of the article in Figure 8.2 is:

  The standard error (SE) of a statistic (most commonly the mean) is the standard deviation of its sampling distribution, or sometimes an estimate of that standard deviation.

  Standard error is distinct from *estimated standard error*. Revise this first sentence to note this distinction rather than simply say that "standard error" has two different meanings. Consider breaking the sentence into two to make this distinction clear.

- Further revise the first sentence to remove the ambiguous use of "its".

- The second paragraph goes into detail about the standard error of the mean. The definition provided holds only for simple random samples, but this qualification is not mentioned. Add a mention of this restriction in your revision.

- The second paragraph also explains why a statistic has a sampling distribution. Some of this information describes general properties of the standard error of a statistic, and it would be more accurate and informative to not limit the description to the special case of the mean. Reorder the sentences so they first address general properties held by a standard error of any statistic, before defining the SE of the mean.

- Rewrite the third paragraph to refer to the standard error of the fitted coefficients, rather than the standard error of the regression.

- The third paragraph contains links to references and other Wikipedia pages. Double-check the referenced sources to make sure they are relevant. Consider adding an additional reference that you trust and think appropriate.

9 Crafting Words and Sentences

General advice for strengthening the details in science writing, such as how to trim phrases, write in a straightforward manner, and use an active voice and concrete nouns, is illustrated via examples in this chapter. In later sections, we examine writing sentences from a higher level, balancing between specific information and general statements, and smoothly transitioning between paragraphs and sections to improve the reading experience and provide a road map for the reader. Finally, we provide a brief list of common grammatical errors to watch out for. Chapter 10 will provide strategies for diagnosing deviations from this advice in a first draft.

9.1 Straightforward Sentences

Scientific writing aims to be precise and concise. There is no place for extraneous phrases that add little or no information. We also avoid strings of modifiers and strive to write plainly and clearly. Removing and trimming phrases and modifiers tightens our writing so the reader can focus on the important parts. When we use plain, clear English, our writing is not necessarily boring or overly simple, but we must pay attention to sentence structure because variation in structure can maintain the reader's interest.

9.1.1 Eliminate Empty Phrases

Phrases that contain no information, such as *it is interesting to note that, the fact that*, and *it should be pointed out that*, should be eliminated. If something wasn't interesting then we would not be writing about it. If something is more interesting than other things, then we can find a more compelling way to draw attention to it. Particularly offensive empty phrases are pompous ones, such as *as is well known, of course, clearly demonstrate*, and *it is obvious*. These are unnecessary and annoying to the reader.

Communicating with Data: The Art of Writing for Data Science. Deborah Nolan and Sara Stoudt, Oxford University Press (2021). © Deborah Nolan and Sara Stoudt. DOI: 10.1093/oso/9780198862741.003.0009

Example: *Empty Opener*
The following sentence explains the reasons for analyzing a subset of the available data. It takes a cumbersome approach that includes unnecessary phrases, redundant information, and clunky descriptors.

In this part of our analysis, we assume that flight delays that last shorter than 15 minutes have minimal effects on passengers, and so we reduce our large dataset into a smaller subset in which all departure delays are at least fifteen minutes long.

A more streamlined version is

Since short departure delays have minimal impact on travelers, we analyzed only those flights where the delay was longer than 15 minutes.

The opening phrase *In this part of our analysis* does not contain information so we dropped it. The adjectives *large* and *smaller* don't add information, and the process of reducing the dataset doesn't need to be described. The essential information is that a subset of the data was analyzed. Notice too that we replaced the clunky description *last shorter than 15 minutes* with *longer than 15 minutes* and mentioned the criteria for subsetting only once. ∎

Example: *Identifying Facts*
A researcher discovered a relationship holds across different groups and writes:

Another point is the fact that all four lines are increasing in average time, regardless of age group.

The phrase *the fact that* dilutes the sentence, and the phrase *another point is* is unnecessary. Without them, the sentence gets to the point quickly and makes a strong statement.

All four lines increase in average time, regardless of age group.

We have also used active voice in the revision to make the statement stronger. ∎

Example: *True That*
In the following sentence, a researcher describes his finding as something that is true.

It is true that websites with a higher number of updates tended to be updated regularly.

Presumably, the researcher would not write about a finding that isn't true so we remove the empty phrase *it is true that*. The sentence is stronger without this phrase.

Websites that were updated frequently within a month tended to have updates at regular intervals.

The revision clarifies the difference between types of websites (those with more or fewer updates) and uses the active verb *have updates,* rather than the passive *be updated.* ∎

9.1.2 Trim Fat Phrases

Fat phrases differ from empty phrases in that they cannot be eliminated entirely. Instead, we often can trim them to a single more direct word. For example, rather than write, *has the ability to,* we use *can.* Instead of *in the event that,* we simply write *if,* and we reduce *the question as to whether* to *whether.*

Example: *Weak Phrases*
An abstract for a grant proposal states:

The proposed research program has the aim to develop theory to bridge the gap between causal inference and machine learning research.

The phrase *has the aim to develop* is weak and indirect. An alternative is to write:

The proposed research will pursue theory to bridge the gap between causal inference and machine learning research.

The new verb *pursue* is active and keeps the necessary ambiguity about the future success of the research.
   This grant abstract concludes with the following statement.

Recruitment and retention efforts with the potential ability to increase the diversity of domestic students is another important goal of the project.

The hedging in the fat phrase *with the potential ability to increase* can be streamlined.

The project emphasizes recruitment and retention efforts to increase the diversity of domestic students.

Using *emphasizes* captures the intent of the statement more powerfully. Dropping the unnecessary claim that this project goal is *important* further focuses the sentence. ∎

9.1.3 Reduce Strings of Modifiers

Too many adjectives and adverbs can muddy an important comparison and dilute the meaning of the modifiers. If all of the modifiers are important, then explain more about them. Expand the sentence to use the adjectives and adverbs in separate clauses so it is clear how they are relevant. Also, consider breaking the sentence into multiple sentences that are easier to read and dropping irrelevant modifiers. We find Alley's analogy helpful in identifying problems with modifiers: consider nouns as stepping stones in a sentence. If strings of modifiers before the noun-stones make it hard for the reader to figure out where to "step", then we rewrite.

Example: *Too Many Modifiers*
The following sentence (from Alley) has nine modifiers for *plant*.

Solar One is a 10-megawatt solar thermal electric central receiver Barstow power pilot plant.

In this sentence it is particularly difficult to discern what *solar thermal electric central receiver* means. Below is a revision that breaks the sentence into two and places the adjectives within clauses.

Solar One is a solar-powered pilot plant located near Barstow, California. Solar One produces 10 megawatts of electric power by capturing solar energy in a central receiver design.

These two sentences make clear that Barstow refers to the location of the plant, the plant uses solar energy to produce 10 megawatts of electric power, and the plant has a central-receiver design. ∎

9.1.4 Avoid Clichés and Colloquialisms

Clichés and colloquialisms include overused sayings that have lost their original meaning and informal terms used in private conversation. Extreme examples are: *eat my dust, kick the bucket, pass the buck, come up to speed,* and *walk on egg shells.* These phrases don't elucidate and can be difficult to understand for those whom English is not their first language.

Example: *Confusing Colloquialism*
In the following statement, we can imagine that someone might take the word *lens* to mean some sort of filter was applied to the data and wonder how the fitting was done.

We fit the distribution through the lens of a quantile-quantile plot.

The phrase *through the lens of* is unnecessary and confuses; we can replace it by *with* as shown here.

We fit the distribution with a quantile-quantile plot.

Another possibility is to reword the sentence with a more precise description of the role of the quantile-quantile plot in the fitting process.

We visually examined the fit of the distribution with a quantile-quantile plot.

Both alternatives make it clear that the quantile-quantile plot was used to assess the fit of the distribution. ■

Example: *Overused Saying*
In the following sentence, the author wants the reader to know that they discovered something unexpected.

Another interesting point that caught my attention is that all four lines are increasing in average time, regardless of age group.

The author would not be making this observation if they didn't think it was interesting so there is no need to tell us that it *caught* their attention. We can drop the phrase entirely from the sentence.

Another point is that all four lines increase in average time, regardless of age group.

Notice that we also dropped the adjective *interesting* because any point we are writing about should be interesting, and we changed the verb from passive to active. Depending on the context, we might also consider dropping the phrase *Another point is that.* ■

9.1.5 Vary Sentences

Writing flows better when there is variation in the length and structure of the sentences. While we strive for straightforward sentences, that does not mean that all our sentences must be simple. To figure out how well your writing flows, try reading it out loud (see Section 10.1 for more on reading out loud). A first-order check should be whether your writing makes sense, then, consider the flow by answering the diagnostic questions:

- Does my writing sound monotonous? (repetitive sentence structure)

- How is the pace? (repetitive sentence length)

- Where do I run out of breath? (run-on sentences or over-complicated wording)

Sentences consist of independent clauses that can be stand-alone sentences and dependent clauses that are sentence fragments. Simple sentences are composed of a single independent clause, compound sentences have two (or more) independent clauses joined by a conjunction, and complex sentences mix an independent clause with one or more dependent clauses. Below are examples of each.

(simple) The plot shows a steady increase in enrollment over the past five years.

(compound) The plot shows a steady increase in enrollment over the past five years, but this year we see a large decline.

(complex) Even though this year's enrollment is low, we see a steady increase in enrollment over the past five years.

Example: *Too Many Simple Sentences*
Note the lack of variability in sentence structure in the following paragraph.

Glyph-maps enable the study of temporal patterns in multivariate spatiotemporal data. Climate change is focused on changes over time. Glyph-maps provide a way to explore these changes directly. Glyph-maps enable different resolutions of the data to be examined. These different resolutions are the raw data to discover data quality issues, global trend, seasonality, residuals, and multivariate dependence. The glyphs are organized in a manner that makes reasonable comparisons when the data is spatially gridded. The icon size needs to be chosen to minimize overlap for irregularly gridded data.

Nearly all of the sentences are simple. In contrast, the following text has greater variation and makes for a less clunky read.

In summary, glyph-maps enable the study of temporal patterns in multivariate spatiotemporal data. Climate change is focused on changes over time so glyph-maps provide a way to explore these changes directly, and they enable different resolutions of the data to be examined: the raw data to discover data quality issues, global trend, seasonality, residuals, and multivariate dependence. When the data is spatially gridded the glyphs are organized in a manner that makes reasonable comparisons. For irregularly gridded data, the icon size needs to be chosen to minimize overlap, but uses as much display space as possible, or combined to produce close to gridded icons. ∎

9.1.6 Straighten Convoluted Sentence Structure

Unusually complex sentences can be confusing for the reader. Such sentences often have too many prepositional phrases. They should be broken up into multiple sentences and rearranged so that they place the desired emphasis on the important ideas. Avoiding particularly long sentences is preferable, but the length of a sentence does not necessarily indicate whether or not the sentence is convoluted. Long sentences with clear flow from one noun to the next are fine, especially if they add variety to the sentence lengths.

Example: *Confusing Clauses*
The following sentence describes how the data were split into training and test sets.

First of all, we divided the dataset into training and test data after shuffling the whole data, which led to 200 training data and 54 test data.

The sentence is hard to read with its two ending phrases. The sentence also begins with an empty phrase that we can eliminate without loss of information. An alternative might be:

We divided the data at random into test and training sets of 54 and 200 observations, respectively.

This shorter statement is clearer and no information is lost. ■

Example: *Awkward Clause Ordering*
The sentence below contains a lot of information about landing patterns at San Francisco International Airport (SFO) airport.

When visibility is low, the Federal Aviation Administration allows airliners to land side by side if runways are at least 4,300 feet apart, but airlines have to arrive one at a time on foggy days since the distance between SFO's runways is only 750 feet, cutting SFO's arrival rate in half to about 30 planes an hour.

All of these ideas have led to a convoluted sentence. It is helpful to know the Federal Aviation Administration (FAA) requirements, but the placement in the sentence is awkward. Splitting this sentence into three helps.

In the fog, the arrival rate at SFO is cut in half to 30 planes an hour. The slow down occurs because the runways are only 750 feet apart so planes can't land side by side. (The Federal Aviation Administration requires runways to be at least 4300 feet apart for planes to land side by side in low visibility conditions.)

We state the most important idea first—the problem with fog at SFO. After that, we explain that the reason for the slowdown is due to the distance between the runways, and, lastly, we provide the FAA regulation as a parenthetical remark. The reordering changes the emphasis of the sentence and reads more smoothly. ■

9.2 Choosing the Right Word

Good technical writing is direct, concise, and precise, requiring attention to the choice of individual words. One rule of thumb is to use concrete nouns and strong verbs. Alley's analogy, made earlier in this chapter, compared nouns to the stepping stones in a sentence. Concrete nouns clearly identify these stepping stones, whereas nouns such as *it* do not. Continuing with this analogy, Alley notes that the verbs in the sentence provide the momentum for moving from one stepping stone to the next, and active verbs provide greater momentum than passive ones.

9.2.1 Use Concrete Nouns

The pronoun *it* often makes the reader pause and ask what is *it*? If there is any doubt about *it* or the reader needs time to figure *it* out, then replace *it* with a concrete noun. Similarly, *this* is a directive that when used as a pronoun creates confusion. If you have written *this* as a pronoun, ask yourself *this what*? The answer guides a revision that uses *this* as a directive. More generally, the nouns we write should be precise. A few common examples of imprecise nouns that crop up in technical writing are *approach*, *process*, and *finding*. Our writing becomes more informative if we replace these nouns like *it* and *this* with more precise ones.

***Example:** Imprecise Pronoun*
We demonstrate the question-and-answer process to make *this* more concrete. Suppose we have written the sentence:

This is efficient.

We ask: this what? Say, our answer is: this algorithm. Then, we revise the sentence to:

This algorithm is efficient.

To continue with this example, suppose we had instead revised our sentence to:

This approach is efficient.

Now, ambiguity still remains about the subject of the sentence. The word *algorithm* more precisely states what is efficient. ∎

9.2.2 Use Strong Verbs

One piece of classical advice for technical writing says to avoid using the first person because it can be perceived as unprofessional to inject ourselves into objective, scientific work. However, avoiding *I* and *we* often leads to passive voice. We recommend judicious use of *I* and *we*. We also advise limiting the frequency of sentences that start with *I* and *we* because the practice of placing *I* at the beginning of a sentence places undue emphasis on the author. Furthermore, take care to avoid fat *I*-phrases, e.g., "I made the choice" can be trimmed to the stronger statement *I chose*.

***Example:** Fat I-Phrases*
Fat phrases can creep into our writing when we talk too much about the process of our analysis rather than the analysis itself (see Section 9.4.1).

Since I saw that the relationship was nonlinear, I made the decision to use a transform.

A more compact version follows.

Since the relationship was nonlinear, I used a transform.

Similarly, a sentence that contains the fat phrase *was able to* to talk about findings becomes bulky.

Thanks to my model's output I was able to determine that there is a significant relationship between a mother' smoking behavior and their baby's weight.

This can be streamlined as follows:

The model output showed a significant relationship between a mother's smoking behavior and her baby's weight. ■

Example: *Passive and Weak Verbs*
The following data description has several occurrences of passive voice, e.g., *were catalogued* and *were provided*, and weak verbs, e.g., *consist of* and *represents an*.

The wildfire data in this analysis were cataloged and compiled by the US Geological Survey, Western Ecological Research Center, and were provided to us by CalFire. They consist of wildfire occurrences on CalFire protected areas from 1 January 1990 to 31 December 2006. The dataset consists of several thousand wildfires and represents an accumulation of a wealth of wildfire data collected from an assortment of different sources.

The revision uses active verbs and a more straightforward wording.

CalFire provided us with wildfire data collected by the US Geological Survey, Western Ecological Research Center. Each record reports information about a wildfire on a CalFire protected area between 1 January 1990 and 31 December 2006. The dataset contains several thousand wildfires, collected from an assortment of different sources.

These more direct verbs help the reader focus on the main points in the paragraph. ■

9.2.3 Match a Word's Connotation with the Context

A word with a negative connotation should not be used in a positive or neutral setting (and vice versa). For example, the word *adequate* in the context of *adequate predictor* implies the predictor is insufficient because we want predictors to be highly accurate.

Example: *Mismatched-Word Connotation and Context*
Consider the following sentence about the sampling process for a survey.

Self-selection of respondents enhances bias.

Bias is an undesirable property, and *enhance* has a positive connotation. They don't fit together. The verb *increases* is a better choice because it does not have a positively associated meaning.

Self-selection of respondents increases bias.

This alternative also conveys the problem with self-selection more precisely. ■

9.2.4 Take Care Swapping in Synonyms

Words with similar meanings are typically not perfectly interchangeable. At times we are concerned about repeating a word too often or too close to its previous occurrence so we search for substitutes. However, substitutions can lead to imprecise and even incorrect statements. It is acceptable to repeat a word, if it is the precise word needed. In the case of mathematical terms, synonyms should not be used.

Example: *Imprecise Synonyms*
Suppose that we want to replace the noun *equal* in the following sentence with a synonym. For fun, we continue replacing the new noun with a synonym for it.

He treated me as an equal.
He treated me as a peer.
He treated me as a colleague.
He treated me as an associate.

Being treated as an equal is not the same as being treated as an associate. This example demonstrates how a sentence can lose precision when we get carried away with synonyms. ■

9.2.5 Avoid Overly Complex Words

Many words that end with *ize* or *-ization* or *-ability* are unnecessarily complex. Examples are the noun *prioritization*, the verb *utilize*, and the adjective *discretized*. Simpler substitutes might be *priority*, *use*, and *discrete*. We recommend tracking complex words that you have a habit of using and challenging yourself to find simpler substitutes. A smattering of complex words will likely have little effect on your writing, but they can quickly combine to be impenetrable.

Example: *Complex Words*
The following sentence contains several complex words that are a mouthful to read, including the aforementioned *utilize*.

The actualization of applied research in statistics must utilize a disinterestedness of the researcher.

What is the sentence saying? Below is what we think the point is:

In practice, good applied statistics requires the researcher to act for the benefit of a common scientific purpose, not for personal gain.

This sentence may be longer than the original, but the intent is clearer. ∎

9.2.6 Eliminate Redundant Adjectives

It is easy to fall into the trap of using redundant adjectives and adverbs when trying to make a strong argument. To keep the writing concise, eliminate these redundancies. Some typical pitfalls and suggested revisions are below.

already existing—No need for already if something is existing.
introduced a new—Here new is redundant with introduced.
mixed together—Mixing implies together.
had done previously—The past tense had makes previously unnecessary.

Example: *Redundancies*
In the following description the adjective *smaller* is redundant in the context of data reduction.

We reduce our large dataset to a smaller subset in which all departure delays are at least fifteen minutes long.

The adjective *large* might also be unnecessary, if the original size of the dataset is irrelevant.

We reduce our dataset to only those flights with departure delays longer than fifteen minutes.

We dropped *subset* too and only describe the selection criteria for the subset. ∎
    We can go too far with streamlining our writing and end up with an overly terse writing style. A little redundancy can help with the flow, and can be more informal. As long as we are aware of redundancies, then we can choose to use them on occasion.

9.2.7 Place Words and Phrases Mindfully

The location of a word or phrase in a sentence can create ambiguities in the meaning of the sentence. The words *only* and *well* can be particularly problematic because they can be used both as adjectives and adverbs.

Example: *Misplacement of Only*
The following five sentences are identical except for the placement of the word *only*.

Only I measured the width of the female crabs after molting.
I only measured the width of the female crabs after molting.
I measured only the width of the female crabs after molting.
I measured the width of only the female crabs after molting.
I measured the width of the female crabs only after molting.

The first instance of *only* refers to the author and indicates no one else took measurements; the second implies the crabs were not subjected to other sorts of treatment; the third implies that other types of measurements (e.g., circumference, weight) were not taken; the fourth suggests male crabs were not measured; and the last occurrence indicates that no measurements were taken before the crabs molted. ∎

Example: *Ambiguous Phrases*
In the following sentence, the introductory phrase is ambiguous.

In foggy weather, the runways are too close for the planes to land side-by-side.

The placement of *In foggy weather* at the beginning of the sentence implies the runways move depending of the weather. As an alternative, we revise the sentence to:

The runways are too close for the planes to land side-by-side in foggy weather.

When we place the modifying phrase at the end of the sentence, closer to *planes*, the sentence is clearer. ∎

9.3 Grammatical Details

Grammatical errors distract the reader from our message. Content and writing construction are entangled, so we recommend assuming that whoever reads your paper will be evaluating the grammar, sentence structure, and style, as well as content. Collected here are a few common grammatical errors, with examples.

- Disagreement of subject and verb or noun and pronoun.

Theories of community ecology suggests that the relationship between species is multi-dimensional.

Theories is plural so the verb should be *suggest*.

Everyone involved in the study can bring their own reading material to pass the time.

Everyone is a mass noun so the pronoun *their* is incorrect. Instead, replace *Everyone* with *Participants*.

- Flawed comparison.

An extra covariate in the model makes it better.

This is an incomplete comparison. We want to know *better* than what?

The people in some communities are much more generous than my neighborhood.

The people are the subject of the sentence and they need to be compared to other people, not other communities. Insert *those* into the comparison, i.e., *than those in my neighborhood.*

- Parallel structure.

In the local marathon, experienced runners start the race slowly and first-timers are starting more quickly.

The experienced runners are described using active voice, but the first-timers are described in the passive voice. Change the description of the first-timers to the active voice *and first-timers start more quickly.*

The statistical bootstrap may be used to compute a confidence interval and as a quantification of the uncertainty.

The two uses of the bootstrap need parallel structure; that is, they need to match, e.g., *to compute* and *to quantify.*

- Too much passive voice.

Lung cancer and other respiratory diseases can be caused by smoking.

More often than not, it is preferable to say *Smoking causes.*

Statistics is still often considered part of mathematics.

An alternative that uses active voice is *People still often consider.*

- Sentence fragment.

The algorithm, obscure in the '90s (although some evidence of use has been found) rapidly becoming the main approach for fitting the model.

This sentence has no verb. Add *is* before *rapidly* to make this fragment a sentence.

That statistical test may be used. But only when the data is normally distributed.

These two sentences should be combined into one because the second is a fragment.

- Wrong tense and/or part of speech.

Some researchers feel badly about how lab animals are treated.

The word *badly* is an adverb and implies that the subject of the sentence has trouble feeling. Change *badly* to the adjective *bad*, i.e., *feel bad.*

Ever since the mayor's office requested an analysis of his new policy, we investigated several options.

The term *Ever since* implies the topic of sentence describes the current situation so use the present tense, *we have investigated.*

- Misused preposition.

My class partner insisted to do the project their way.

The proposition *on* rather than *to* is required: *insisted on doing the project.*

- Wrong placement of word or phrase.

When applying an ordinary least squares regression, the resulting model describes the linear relationship between two variables.

The first phrase in the sentence is dangling. It should be *When one applies.*

Cloud computing can produce byproducts that are bad for the environment such as carbon emissions.

The phrase *such as carbon emissions* is misplaced and should be moved earlier to *can produce byproducts, such as carbon emissions, that are bad for the environment.*

- Illogical connection among subject/verb/complements.

Viewing residual plots suggests that the model fits poorly.

It is not the viewing, but the residual plots themselves, that suggest the misfit. Drop the introductory phrase about viewing and make the plots the subject of the sentence, e.g., *The residual plots suggest.*

Exposure to intense x-rays is the reason these crystals glow.

The exposure itself is not the reason the crystals glow. The exposure causes the glowing, e.g., *Exposure to intense x-rays causes these crystals to grow.*

- Wrong or vague reference.

The updated data allow us to examine trends in poverty over time. It's a great breakthrough.

The *it* in the second sentence is a vague reference. Make *it* concrete with, e.g., *Revision of the data collection process is a great breakthrough.*

- Run-together sentence.

The logistic regression is the most apt model at the moment, however, other approaches do exist.

Two sentences run into one. This problem can be fixed by replacing the comma with a semicolon or by splitting the sentence into two, i.e., placing a period before *however*.

The dataset is fairly complete, it has only a few missing values.

The comma should be replaced with a semicolon or period, or a conjunction such as *and* should be inserted after the comma to create a compound sentence.

9.4 Describing Your Findings

Well-written technical reports present arguments based on data in a logical and convincing manner. Beyond the standard writing techniques of Sections 9.1, 9.2, and 9.3, good writing presents results clearly and directly without editorializing about the process of obtaining them, includes information in a logical order that the reader can easily navigate, and limits the use of technical jargon. Good science writing also aims to streamline the reading process by providing helpful transitions between ideas.

Writing well, while maintaining the interest of the reader, is an art that requires plenty of practice. Being a careful reader helps build intuition for what smooth and awkward writing looks like. Pay attention to how writers you admire build their paragraphs with different sentence lengths and structures and how they transition between paragraphs.

9.4.1 Tell What You Found, Not the Path You Traveled

For beginning scientists, one of the hardest skills to learn about writing is to avoid presenting findings in the order in which the analysis was carried out. There are a few circumstances where we do want to write about the process in chronological order, for example, in a blog post that demonstrates the thought process behind an analysis, but most often we want to present a summary of our findings not a description of the analysis process. For example, if we fitted several alternative models to the data, then we typically mention these models in a few sentences in a discussion section and do not dedicate space to describing these alternatives in the sequence that we tried them. As another example, exploratory data analysis is an important and often lengthy stage of the analysis that can uncover problems with the data, the need for transformations, and unexpected relationships to consider when modeling. It is important for replicability that we describe these findings and make our code available so that the reader knows exactly what we did to the data and whether there was data snooping, but this description typically constitutes a brief summary in our written report. More information about replicable code can be found in Section 5.5.

Example: *Describing the EDA Journey*
The following paragraph justifies the researcher's transformation of the data.

During my exploratory data analysis, I found that the distribution of house prices is heavily right tailed. Hence, I applied a log transform to housing price and achieved a relatively normal distribution.

The relevant point is that the log-transformed data have an approximate normal distribution.

The log transformation of house price follows an approximate normal distribution, so we analyze the transformed data.

We trimmed the phrase *during my exploratory data analysis* and the description of the untransformed data. If the distribution of the log-transformed data looks normal, then the distribution of the untransformed data must be right-skewed. ∎

Example: *Inserting "I" in the Findings*
The first sentence below in the description of a distribution adds no information about the findings. This sentence simply states that the researcher examined the income distribution.

We also look at the income level of our survey participants by the annual household income in thousands from all sources. The distribution seems to be left-tailed, and the majority of our participants is earning above $50,000 per year.

We eliminate the first sentence entirely and edit the second sentence for a more direct description of the relevant findings.

The distribution of annual household income has a long left tail, with the majority of respondents earning over $50,000.

The revised second (and now only) sentence uses a stronger verb than *seems to be*, replaces *above* with the more appropriate comparison *over*, and eliminates *per year* which is implied by *annual*. ∎

Example: *The Specific After the General*
The first two sentences below jump right into specific rates without context. The reader doesn't know what is considered low birthweight, why 1970 and 2014 are being compared, or if the decline is practically significant. The third sentence then summarizes the implication of the two facts, implying that the decline is actually notable.

Roughly 6.2% of singleton births in 2014 were low birthweight. In 1970 7.9% of singleton births were low birthweight. The rate of low-birthweight babies has declined from 1970 to today.

The following description begins with a general statement about the change in the percentage of low-birthweight babies. The second sentence then provides specific details about the actual percentages being compared, and the third sentence provides a benchmark value.

The rate of low birthweight babies has declined from 1970 to today. Roughly 6.2% of singleton births in 2014 are low birthweight, compared to 7.9% in 1970. Our analysis uses National Center for Health's definition of low birthweight of under 2500 grams (5.5 pounds).

These percentages are based on all births in the USA so there is no need to provide standard errors. ∎

Example: *Accuracy Statements Too Early*
The first sentence gives the reader a specific measure of performance. However, the reader doesn't have an anchoring value to determine whether or not this implies reasonable accuracy. The second sentence asserts that this number means that the approach is indeed accurate, but does not give further evidence to support the claim.

The nearest neighbor method of geolocation predicts the location to within 8 ft on average. This approach is accurate for the test data.

Below, the prediction method is first described generally as accurate, followed by a more specific quantification of the average distance between the prediction and the truth.

The nearest neighbor method accurately predicts the location of the test data. The proposed method of geolocation predicts the location to within 8 ft on average, with a standard deviation of 2 ft.

In addition to the average distance, the second sentence provides a measure of the typical deviations between the predicted and actual location. ∎

9.4.2 Provide Helpful Transitions

When we think of transitions, obvious words, such as *next, finally, however,* and *then,* come to mind. Relying on these simple transitions to smooth the flow between ideas in our writing can lead to a clunky writing style. We can afford to be more subtle while still connecting our ideas. The goal of an effective transition is to facilitate easy reading, and although we want to avoid transitions that are too general or too specific, we also want to avoid filler transitions that add no content.

In more formal technical writing, readers often jump around instead of reading a paper straight through. The use of transitions should anticipate this and help individual paragraphs contain enough context to stand alone. Additionally, section and subsection headers can act as transitions and allow a reader to skim the piece at a high level and gain a quick road map. In this case, section headings may negate the need for transition text at the beginning and ending of a paragraph.

Example: *Jumping in Quickly*
The first paragraph below ends with a question as a transition, and the second paragraph begins with an answer to that question.

...We fitted our models using least-squares regression. But which model do we prefer?

For drugs such as wormers and antibiotics the therapeutic window is quite wide, and it is better to overdose the donkey because otherwise the infestation/infection might not be treated.

The authors smoothed this transition with a few sentences that frame the choice of methodology in terms of the original problem.

...We fitted our models using least-squares regression. But which model do we prefer?

Consider the loss function from the point of view of the donkey's health. This loss function is the cost of getting it wrong. It depends, among other things, on the drug that is being prescribed. For drugs such as wormers and antibiotics the therapeutic window is quite wide, and it is better to overdose the donkey because otherwise the infestation/infection might not be treated.

These additional introductory sentences recall the purpose of the study and add support for the researchers argument for a non-traditional loss function. ∎

Example: *Lack of Positioning*
An article on relative age effect (RAE) first considers RAE in athletics, and then moves to consider the impact on academic achievement. One section begins as follows:

Does the RAE occur in more important areas that have long-term ramifications? Musch and Grondin linked the RAE to negative consequences for personal development of those who are the youngest in their cohorts in youth sport programmes.

This introductory paragraph jumps right to the relevant question. The following version inserts a transition paragraph that summarizes the findings about RAE in athletics and places these findings in a larger context.

The evidence of RAE in youth sports is compelling; but the inability of a child to compete on a level playing field in world football, hockey, or baseball hardly seems too tragic. For most of us, our sporting pursuits eventually become recreational hobbies or are abandoned as we move into adulthood. This leads to an important question: does the RAE occur in more important areas that have long-term ramifications for the rest of us?

In addition to the transition to a *bigger* question, this paragraph provides context clues about RAE findings in athletics. This extra context helps a reader jumping into the article at this point. ∎

Example: *Extraneous Transitions*
We can get carried away and construct transitions from unnecessary interesting information. The following is an extreme example, where we inserted details from Wikipedia for the transition.

The Edwin Smith Surgical Papyrus, written in the 17th century BC, contains the earliest recorded reference to the brain. In 1997, a research team headed by Streissguth launched a detailed study of the corpus callosum in MR images of the brains of 180 people.

This transition bears little on the research conducted; the additional information does not help the reader understand the motivation or background for the study. Below is an alternative that connects more directly to the research.

Stimulated by the pioneering report of Riley et al. (1995), in 1997 a research team headed by Ann P. Streissguth launched a detailed study of the corpus callosum in MR images of the brains of 180 people.

This transition sentence provides motivational context for the research, including a reference to earlier work that contains useful background information. ∎

Example: *Lack of a Thought-Provoking Transition*
The paragraph below begins with a clear statement of the researchers' findings, which are qualities that we want our writing to embody.

Our analysis suggests that Barham's policy of abandoning the transatlantic slave trade early could have led to more female slaves being allocated to the field gangs, thereby worsening their survival prospects.

However, the more engaging transition below tells the reader that the analysis found a counterintuitive result.

One might suppose that Joseph Foster Barham II's policy of amelioration would have eased the lot of his slaves and prolonged their lives. However, our analysis suggests that his policy of abandoning the transatlantic slave trade early could have led to more female slaves being allocated to the field gangs, thereby worsening their survival prospects.

This transition pushes the reader to consider two alternative arguments. The obvious supposition may not be necessary, but the contrast captures the reader's attention. ■

9.5 Breaking Some Traditional Writing Rules as a Blogger

Decisions made at the word and sentence level will differ between informal and formal writing. One extreme case of informal writing that gives the writer more flexibility is the blog. A reader of a blog post not only wants scientific content in a more accessible format, but they also want to get a sense of the writer's personality, the scientist behind the science. Therefore, some of the rules described earlier in this chapter may be broken in service of the differing goals of a blog post.

The following examples contrast the informal blogging style that breaks many of our earlier principles with a more formal style that adheres to these rules.

Example: *Fat Phrases*
We have advised against using extraneous phrases that add little to no information about the results in a formal report. The excerpt below stays to the point.

There are nine circled fouls where game results might have been affected with the redesign. What would have happened with those fouls if the penalty box had not been there? They would still need to be penalized in some way, just not with a penalty kick. Fouls in the shaded regions could become free kicks instead of penalty kicks.

However, in a blog post phrases such as *it is interesting to note that* or abrupt interjections give the reader information about the writer, if not the scientific content. "Fat phrases" that add insight into the writer's thought process, such as *As we can see* and *The first thing that comes to mind*, appear in the blog excerpt below.

As we can see, there are nine circled fouls where game results might have been affected with our redesign. But what exactly would have happened with those fouls if the penalty box had not been there? They would still need to be penalized in some way, just not with a penalty kick. The first thing that comes to mind is that fouls in the shaded regions would become free kicks instead of penalty kicks. ■

Example: *Process*
Information about the data-analysis process does not add information about the results. The excerpt below has no mentions of this process.

Data available include start dates for each season of the bachelor franchise, ratings for each season, and additional metadata about the show.

In contrast, the following blog excerpt tells us about the inner workings of the writer's mind and gives us a sense for the writer's plan of attack for the analysis

We are going to bring in a dataset which has the start dates for every single season of the bachelor franchise. We will then do some data munging to find the closest ratings date to the season start date. With this info we will join the bachelor season metadata to the ratings table. ∎

Example: *Informal Terms*
Informal terms used in private conversation do not match a professional tone. The following sentence avoids informalities.

To determine how many clusters to assume for the hierarchical clustering algorithm, the NbClust package and its documentation was consulted.

However, blog posts often use informal language to make the prose more accessible. The colloquial phrases *rabbit hole* and *silver bullet* make the text in the blog excerpt below more conversational.

This post will be taking a bit of an unexpected diversion. As I was experimenting with hierarchical clustering I ran into the issue of how many clusters to assume. From that point I went deep into the rabbit hole and found out some really useful stuff that I wish I'd have known when I wrote my previous post. ...
Could I have found my silver bullet?
Eager to try it out for myself, I discovered the NbClust package and read its documentation. ∎

Example: *Using I*
I is often deemed too informal for a report, but as described in Section 9.2.2, the use of *I* can make it easier to use active and concrete words, including strong verbs. The following paragraph avoids *I*.

The games were normalized by length so that those that go into overtime are not automatically chosen as "exciting." Games, teams, and conferences were explored using the ncaahoopR package.

It is common to use *I* in blog posts, e.g., the phrases *were normalized* and *were explored* above are avoided in the informal excerpt below.

The reason I choose to normalize the length of games is that I don't want sloppy "boring" games which simply happen to go to 2 or 3 overtimes to be pegged as more exciting. In general, this small normalization has little effect, as games that go deep into overtime are generally pretty exciting to begin with. Through this article, I hope to explore which games, teams, and conferences have produced the most exciting basketball this season, while showing off how one can use ncaahoopR to answer interesting college basketball questions. ■

Example: *Informal Transitions*
Sentence variation remains crucial to maintain momentum for the reader, whether in formal or informal writing. However, a blogger, unlike a report writer, can use strategies such as starting a sentence with lead-in words such as *and*, *but*, and *so* to more casually transition between ideas. Compare the following two passages. The second, more informal paragraph has a simple sentence, a compound sentence, and a sentence that starts with *but*.

Long sequences of all heads or all tails don't look like our conception of randomness. Strings of heads or tails appear overly structured and thereby not random despite long sequences of heads and tails being quite probable under independent coin flipping.

The gist is that human intuition is bad at solving this problem. Long sequences of all heads or all tails don't look like our conception of randomness; they appear overly structured and thereby not random. But long sequences of heads and tails are in fact quite probable under independent coin flipping! ■

By combining these rule-breaking elements, a blog gains a distinctive style marked by conversational language and the writer's role as a character. We provide examples of this style next.

Conversational Language Many blogs are distinguished by chatty language that reads like a conversation with the author. It is not unusual to see sentences starting with *and*, *but*, or *so* even though these would be deemed unacceptable in more formal settings (see Section 9.4.2). Stream-of-consciousness interjections are also prevalent. By writing more informally, we may automatically avoid needlessly complicated words (see Section 9.2.5). These aspects of blog style can also lead to less formal sentence structure than we are used to from Sections 9.1.5 and 9.1.6, but as long as we maintain clarity, these deviations from our earlier advice are suitable for a blog.

Example: *Stiff Blog Posts*
The following sentences would not be out of place in a formal article. The main points follow one after another without any additional editorializing, and we don't get a sense of the personality of the writer.

This is not a sufficient qualification. Not everything that fits each definition is a part of that field. These also are not a good way of determining someone's role or job title. A role is a matter of focus and experience.

However, this definition can distinguish the three types of work in a descriptivist manner, partitioning jobs according to how people in the field typically do.

In contrast, the version below interjects thoughts within the main text, helping us get to know the author through their stream-of-consciousness. Here *I* is used liberally, and when we read the text out loud, we can picture someone talking to us in a casual setting. Other small stylistic changes, such as the use of contractions, give the text a more informal tone too.

To be clear, this isn't a sufficient qualification: not everything that fits each definition is a part of that field. (A fortune teller makes predictions, but we'd never say that they're doing machine learning!) These also aren't a good way of determining someone's role or job title ("Am I a data scientist?"), which is a matter of focus and experience. (This is true of any job description: I write as part of my job but I'm not a professional writer).

But I think this definition is a useful way to distinguish the three types of work, and to avoid sounding silly when you're talking about it. It's worth noting that I'm taking a descriptivist rather than a prescriptivist approach: I'm not interested in what these terms "should mean", but rather how people in the field typically use them. ∎

Ourselves as a Character The use of conversational language helps us transition away from more guarded writing. Although in formal writing we try to remove ourselves from the content we are discussing, a blog allows us to blur the line between our professional and personal ties to the content. In a blog post, the content may be interjected with personal anecdotes or information. Similarly, unlike in Section 9.4.1 where the emphasis is placed on telling the reader what you found, a blog post may be the venue where you want to insert yourself into the story and talk about the path you traveled. Many blog posts document the process behind work that is more formally explained elsewhere.

Example: *Blogging About the Process*
Three sentences can capture an analysis of music data grouped by different workouts. These sentences describe the findings without any process information.

Tracks associated with certain muscle groups have more missing values than others, so the representativeness of the sample may be a problem. The shoulders track is associated with less pop or dance, and the cool down track is quieter with no electronic music. Future work could study beats per minute data.

The blog version is noticeably longer because intermediate steps and conjectures are added. The reader not only gets to read about the results but also how the author found those results (e.g., what plots were made). The steps are outlined in a stream of consciousness style, interjected with mini-brainstorms the author had as they went through the analysis process.

I first made a figure of the number of missing values i.e. missing cluster by muscle group. Sadly some muscle groups seem to have more missing values, hopefully the sample I do have for them is representative. Note that in total I have 42 releases. And then I looked at the musical genres by muscle group.

So, is this interesting? I'm not exactly sure, ah! It seems that the shoulders track is associated with less pop or dance than the other tracks, maybe because overhead presses demand more rhythm? Cool down does seem more quite with no electronic music (edm and big room) which makes sense. All in all, I was happy to see this plot but would like to have a better categorization of songs and well less missing values. I'm not a good musical analyst yet! Oh and I'd love to get data about beats per minute. Maybe in a next release of this post, by me or one of you dear readers? Oh and if you want music that is more quiet, check out this post of mine about Radio Swiss Classic program! ■

As we embrace the nuances of blog writing, the writing advice in this chapter remains relevant. Straightforward sentences and careful word choice should still be prioritized. Basic grammar still matters as well.

9.6 Notes

In his book, *The Craft of Scientific Writing*, Alley [1] discusses the six goals of language for scientific language: precision, clarity, familiarity, forthrightness, conciseness, and fluidity. These goals also shape our discussion in this chapter. Some material in Sections 9.1 and 9.2 was adapted from similar advice in Alley. For example, the "stepping stone" analogy, the "too many modifiers", "misplacement of *only*", "imprecise synonyms" and "eliminate redundant adjectives" examples, the importance of connotation when using "adequate", the wariness of "needlessly complex words" and "arrogant phrases", and the term "fat phrase" are inspired by examples in his book. The term "empty phrase" is also discussed in Gelman's blog post [7].

Examples in Sections 9.1.1, 9.1.4, and 9.1.6 adapt student work. The example in Section 9.1.2 is from a proposal written by Nolan; the example in Section 9.1.3 adapted text from Schweinberg and Rasband [16]; and the example in Section 9.1.5 adapted text from Wickham et al. [19]. The examples in Sections 9.2.2, 9.2.6, and 9.2.7 are from student work, except for the second example in Section 9.2.2, which is adapted from Baltar et al. [2]. The grammatical details in Section 9.3 are adapted from Tollefson [18].

The examples in Section 9.4.1 are adapted from student work. The transition examples in Section 9.4.2 are adapted from text in (in order): Milner and Rougier [13], Brewer and Cochran [5], details from Wikipedia [20] added to Mardia et al. [12], and Smith and Forster [17]. For more guidance on transitions see Chapter 3 of Alley, which includes discussion of "empty" beginnings that don't contribute information, "in medias res" beginnings that start too specifically, and "Genesis" beginnings that start too broadly. For more guidance on varying sentences see Chapter 9 of Alley which includes discussion of varying the rhythm, length, and structure of sentences.

The blog post examples in Section 9.5 come from (in order): Bertin [4], Ellis [6], Lendrum [9], Benz [3], Hayes [8], Robinson [14], and Salmon [15]. The sample sentences in Section 9.8.1 are adapted from student materials, the text of the press release in Section 9.8.3 was adapted from a UC Berkeley press release by Manke [11], and the blog excerpt in Section 9.8.5 is from Lukito [10].

For more practice, consider working on Portfolio Piece F.

9.7 References

[1] M. Alley. *The Craft of Scientific Writing*. Springer-Verlag, New York, 3rd edition, 1996.

[2] M. Baltar, J. E. Keeley, and F. P. Schoenberg. County-level analysis of the impact of temperature and population increases on California wildfire data. *Environmetrics*, 25(6):397–405, 2014.

[3] L. Benz. Game excitement index: An in-depth exploration. https://lukebenz.com/post/gei/, December 2018.

[4] A. Bertin. Redesigning soccer's penalty box: can statistics help us make the sport more fair? https://ambertin99.wixsite.com/statsbloggr/post/redesigning-soccer-s-penalty-box-can-statistics-help-us-make-the-sport-more-fair, August 2019.

[5] L. M. Brewer and J. J. Cochran. August's child is... favoured by fortune. *Significance*, 10(3):20–24, 2013. https://rss.onlinelibrary.wiley.com/doi/full/10.1111/j.1740-9713.2013.00661.x.

[6] L. Ellis. Analyzing the bachelor franchise ratings with gtrendsR! https://www.littlemissdata.com/blog/bacheloranalysis, October 2019.

[7] A. Gelman. Advice on writing research articles. https://statmodeling.stat.columbia.edu/2009/07/30/advice_on_writi/, April 2018.

[8] A. Hayes. Overfitting: a guided tour. https://www.alexpghayes.com/blog/overfitting-a-guided-tour/, January 2020.

[9] J. Lendrum. Experimenting with hierarchical clustering in a galaxy far far away... https://r-house.netlify.com/2019/07/15/experimenting-with-hierarchical-clustering-in-a-galaxy-far-far-away.../, July 2019.

[10] J. Lukito. Disney plus data and chill ;). https://www.jlukito.com/blog/2019/11/22/disney-plus-data-and-chill-, November 2019.

[11] K. Manke. Genomic gymnastics help sorghum plant survive drought. Media Relations, UC Berkeley, December 2019. https://news.berkeley.edu/2019/12/02/genomic-gymnastics-help-sorghum-plant-survive-drought/.

[12] K. V. Mardia, F. Bookstein, and J. Kent. Alcohol, babies and the death penalty: Saving lives by analysing the shape of the brain. *Significance*, 10(3):12–16, 2013. https://rss.onlinelibrary.wiley.com/doi/full/10.1111/j.1740-9713.2013.00659.x.

[13] K. Milner and J. Rougier. How to weigh a donkey in the Kenyan countryside. *Significance*, 11(4):40–43, 2014.

[14] D. Robinson. What's the difference between data science, machine learning, and artificial intelligence? http://varianceexplained.org/r/ds-ml-ai/, Janurary 2018.

[15] M. Salmon. The music of Les Mills Body Pump, with Spotify data. http://www.masalmon.eu/2017/10/01/bodypump/, October 2017.

[16] R. Schweinberg and J. Rasband. Solar one—a 10-megawatt solar thermal central receiver pilot plant project. *Conference on Solar Energy: Technology Status*, 1978.

[17] S. Smith and M. Forster. The hazards of slavery. *Significance*, 8(4):179–182, 2011. https://rss.onlinelibrary.wiley.com/doi/full/10.1111/j.1740-9713.2011.00526.x.

[18] S. P. Tollefson. Encouraging student writing. UC Berkeley, Office of the Assistant Vice Provost, Undergraduate Education, 1988. https://teaching.berkeley.edu/sites/default/files/encouraging.pdf.

[19] H. Wickham, H. Hofmann, C. Wicham, and D. Cood. Glyph-maps for visually exploring temporal patterns in climate data and models. *Environmetrics*, 23(5):382–393, 2012.

[20] Wikipedia. History of neuroscience. https://en.wikipedia.org/wiki/History_of_neuroscience.

[21] Wikipedia. List of statistics articles. https://en.wikipedia.org/wiki/List_of_statistics_articles.

9.8 Activities

9.8.1 Focused Revision at Sentence Level

Collected here are several sentences that need revision. For each sentence consider the following four possible actions as you revise. Each sentence is accompanied by some ideas for fixing it.

- *Trim.* eliminate empty phrases, trim fat phrases, reduce modifiers, and drop redundant adjectives.

- *Straighten.* convert a convoluted sentence into a straightforward one, reorder phrases, and break the sentence into multiple sentences.

- *Emphasis.* order concepts by importance, balance general with specific, and define statistical terms.

- *Word choice.* replace weak nouns with concrete ones, passive verbs with active voice, and match the connotation of words with the context of the sentence.

1. The author describes past research in the distribution of precipitation. Consider changing to an active voice and getting directly to the point.

   The investigations into the daily rainfall distribution are primarily spread throughout three main research areas, namely, (1) stochastic precipitation models, (2) frequency analysis of precipitation, and (3) precipitation trends related to global climate change.

2. The author describes a variety of factors that they have chosen to distinguish spam from good email. Eliminate empty phrases, avoid referring to variable names and types, and straighten the sentence.

> For instance, the number of the lines in an email body provides differences between spam emails and useful emails that can't be ignored; a binary variable, which tells us whether an email having "Re:" in the email subject, shows that when the email in the dataset is a spam, it almost never contains "Re:".
>
> Of these 10 logical variables that were selected from the list of 16 as potential indicators of spam status, they were ordered into our 3 general sections:
> - related to the subject line: isRe, subjectSpamWords
> - related to the body content: isOriginalMessage, isDear, isWrote
> - related to the header/email specifications: messageIdHasNoHostname, containsImages, fromNumericEnd, isPGPsigned, isInReplyTo
>
> which we believe may be useful in interpretation for our model later on.

3. In an observational study, the author examines various relationships between math performance and covariates. Improve the transition, remove the causal connotation, and eliminate the "journey" aspect of the investigation.

> After understanding interaction between peers, I am interested in investigating the impact of video games and televisions on math scores; whether spending more time on entertainment would help students perform better in math.

4. The model the author describes aims to predict the location of a Wi-Fi device using the signals received by the access points. Straighten the sentence describing the model, simplify the language, and correct the description of the model.

> If a device catches the proliferated signals from some access points, which send out Wi-Fi signals, a location control center uses a model to estimate the location of the device.

5. The author makes a recommendation for when to schedule flights based on their analysis. Strengthen the verbs, and trim the phrases.

> According to all the findings regarding seasonal effects on delay, it is reasonable to recommend passengers to reduce their times to take a flight in the summer and winter season.

6. The author describes the source of the data. Order the description by importance, and eliminate unnecessary phrases.

The data are scraped from San Francisco Chronicle weekly. However this feature is provided by the California Resource, a title abstracting company. And all the detailed information are collected from SF Bay Area Counties house transaction records.

7. The author describes features of the State of the Union Addresses. Reduce and eliminate phrases, straighten the description, and consider the order of the information provided.

Although called "address", not all the addresses are delivered spoken. The format of delivery varied across speeches. While all the speeches extreme in length are delivered written, the written speeches tend to be longer in general.

8. The author describes a problem with biased data and argues that the records corresponding to incomplete observations are eliminated from the analysis. Straighten the sentence, eliminate empty phrases, reduce fat phrases, and use the active voice.

Since we are interested in the gaps between each update, if a website is missing a number of checkpoints, then we will not have complete information of a website, which could lead to extremely biased results if we insist to use them. When a website is checked, let's say, fewer than 700 times, we eliminated it from the analysis.

9. The author states that the performance of models were compared with test data and that their chosen model is good. Streamline this sentence, and rewrite to avoid the analysis journey.

These models were run against a test dataset to verify the accuracy of each model against real cases, and in comparing our analysis of each, we have determined a model that would be a strong implementation for a new spam filter and what patterns in variables that may be useful in further investigation of traits of spam emails.

9.8.2 Writing Transitions

Collected here are paragraphs that summarize a variety of findings from the analysis of the Drug Abuse and Warning Network (DAWN) data (see Section 6.7). Integrate these paragraphs into one coherent and compelling story. To do this, choose an ordering for the paragraphs and write transitions between ideas. Remember, transitions are not just filler! When writing your transition sentences, ask yourself if the reader already knows all of the content contained in your transition sentence(s). If so, then the transition is likely "filler". It's okay to use a few generic transition words or phrases, but tally your use of transition words and phrases. Then revise to avoid using the same one more than once in a row between paragraphs. You can also rearrange the sentences within the paragraphs if it helps you transition more easily between paragraphs.

Paragraph A We see that marijuana usage differs by gender and age. Males make up the majority of the marijuana-related visits. Teenagers have a large proportion of marijuana-related visits with a decline after the mid-twenties. The two areas with the largest number of marijuana-related visits are Houston, TX and the Bay Area.

Paragraph B Drug usage declines for those in their fifties and sixties, except for pharmaceutical drug use. Visits to the hospital mostly result in going home when the patients are young, but the percentage of those who go home after the visit steadily declines in older age groups. There is also an increasing trend in "other inpatient" services being needed as age increases; older patients may be less resilient and need more care.

Paragraph C DAWN is a public health surveillance system of more than 250 hospitals evenly distributed across the USA that monitors drug-related visits to the hospital emergency departments (EDs). DAWN allows us to monitor morbidity and mortality trends associated with substance abuse by capturing ED visits for various drugs, such as prescription drugs, dietary supplements, and illegal substances. Each case in the data is from a visit to the ED that involves recent drug use. The dataset contains several indicators for what kind of drugs were taken, what category of abuse/misuse is involved, why the patient was admitted, and where the patient went after leaving the ED, along with demographic information for each patient.

Paragraph D There are three main types of drug usage of teens: pharmaceuticals (antidepressants), non-medical pharmaceuticals (opiates), and illicit drugs (alcohol related). Are teens more susceptible to drug abuse/overdose? The DAWN data shows that 80% of the visits in this group are caused by pharmaceuticals; only 20% are caused by non-medical pharmaceuticals and illicit drugs. In general, not many drug-related hospital visits are due to suicide. Unlike the use of marijuana, the number of suicide attempts does not seem to have any correlation with gender, race, or age. However, there are gender differences in visit behavior among teenagers overall. For example, the majority of overdose visits caused by antidepressants are female.

9.8.3 Press Release Dirty Half-Dozen

Just as we described the "dirty dozen" recommendations for code that deviates from good programming practices, we have created a set of "dirty half-dozen" recommendations for writing bad press releases. Many of these in Table 9.1 violate the principles of clarity and brevity.

Do	(We really mean...)
1. Report everything	Write a press release only if you have something interesting/important to say.
2. Use complex jargon	Use simple, nontechnical wording, where possible.
3. Pepper with acronyms	Limit the use of acronyms, but if needed define them in plain language.
4. Exaggerate findings	Be honest; include compelling quotes.
5. Waffle on	Keep to 300–350 words; leave journalist wanting to know more.
6. Hedge findings	Get straight to the point; avoid verbs like *might be*.

Table 9.1 The Dirty Half-Dozen.

Consider the following press release excerpt that adheres to the dirty half-dozen recommendations. Find and label an example of each. Then rewrite the paragraphs to avoid them.

A new study published this week in the journal PNAS provides the first detailed look at how the plant might survive when its surroundings turn harsh and arid. Plants may be exercising control over its genome—switching some genes on and some genes off at the first sign of water scarcity, and again when water returns.

The massive dataset, collected from 400 samples of sorghum plants grown during 17 weeks in open fields in California's Central Valley, reveals that the plant modulates the expression of a total of 10,727 genes, or more than 40% of its genome, in response to drought stress. Data collection and management involved five systems, three programming languages, and 100 GB of storage. Many of these changes occur within a week of the plant missing a weekly watering or after it is first watered after weeks of no watering.

The data was collected without flaw as part of the EPICON project, a five-year, $12.3 million study into how the sorghum plant is able to survive the stress of drought. The EPICON study is run as a partnership between UC Berkeley researchers and scientists at UC ANR, the Energy Department's JGI and that agency's PNNL.

The data in the current paper show perfectly the plant's transcriptome under both normal conditions and drought conditions over the course of a single growing season. In the future, the team also plans to publish data from the other two years of the experiment, as well as proteomic and metabolomic data.

The following might be useful for decoding acronyms: "Epigenetic Control of Drought Response in Sorghum", *Proceedings of the National Academy of Sciences*, UC Agriculture and Natural Resources, the Joint Genome Institute, Pacific Northwest National Laboratory.

9.8.4 Crafting a Wikipedia Section

Using the following example as a guide, choose a statistics-related page from Wikipedia's list of statistics articles [21]. Revise one or two paragraphs. Consider what you can make more precise and what you can make simpler. Note the words you replaced to make things more precise. Do any match the ones replaced in the example here? Can you revise such that the new version uses fewer words than the original?

The Wikipedia article on standard error contains a subsection on relative standard error (see Figure 9.1). Just as we edited for statistical accuracy in Section 8.1 we can similarly edit this for parsimony and clarity.

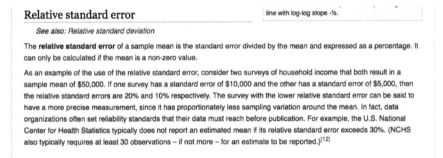

Figure 9.1 Screenshot of the Wikipedia Subsection on Relative Standard Error.

The first sentence of this section defines relative standard error in terms of one quantity divided by another. A more precise statement would simply define relative standard error as a ratio, e.g.,

The relative standard error of a sample mean is the ratio of the standard error to the mean, expressed as a percentage.

The second sentence uses the noun phrase *a non-zero value*, but it may be clearer to the general reader if we describe the quantity simply as *not zero*, e.g.,

It can only be calculated if the mean is not zero.

In the second paragraph, we replace the term *lower* used to describe the 10% case with the more typical descriptor *smaller*. Also in that paragraph, one sentence begins with the empty phrase *In fact*, which we eliminate. In this same sentence, if we modify *set reliability standards* to *set reliability standards in terms of relative standard error*, then we make clearer the connection between the standards and the topic of the section. The revised paragraph appears below.

As an example of the use of the relative standard error, consider two surveys of household income that both result in a sample mean of $50,000. If one survey has a standard error of $10,000 and the other has a standard error of $5,000, then the relative standard errors are 20% and 10%, respectively. The survey with the smaller relative standard error can be said to have a more precise measurement, since it has proportionally less sampling variation around the mean. Data organizations often set reliability standards in terms of the relative standard error that their data must reach before publication. For example, the U.S. National Center for Health Statistics typically does not report an estimated mean if its relative standard error exceeds 30%. (NCHS also typically requires at least 30 observations—if not more—for an estimate to be reported.)

9.8.5 Removing Blog Elements

In Section 9.5, we saw how a blogging style can transform formal prose. In this activity, you will do the opposite, seeing how removing elements of the blogging style affects the text. First, highlight all elements of the blogging style in the following blog excerpt.

On December 12, 2019, Disney unveiled its streaming service, Disney+, to the world. It received significant attention, both good and back, from the press—which makes sense, because over 10 million people signed up in the first day.

Twitter was also abuzz with conversations about Disney+ (see this string-of-tweet "news story" about Twitter activity on the first day). Several pointed out that shows, including new ones like The Mandalorian and oldies like Darkwing Duck, were trending soon after Disney+ was launched.

But what would activity look like after the first day?

To answer this question, I used Mike Kearney's rtweet package to look at tweets posted from 11/14/19 to 11/18/19 that had one of the following keywords: disneyplus, disney plus, disney+, and disney +. ...

I also ran an LDA topic modeling, which highlights the variety of conversations on Twitter about Disney+.

Noticeably, The Mandalorian, Hannah Montana, the Simpsons (which is on Disney+ in its original 4:3 format), and Bad Girls Club were talked about frequently enough to be (mostly) stand-alone topics. The Mandalorian hashtag (#themandalorian) was also a popular keyword in the corpus.

But we also see a variety of other topics, including one about the Nickelodeon and Netflix deal (which many people viewed as a response to Disney+'s explosive popularity) and another comparing Disney+ to other streaming services (like Netflix, Hulu, and HBO). In fact, Netflix was the third most frequent term in the dataset (behind Disney and Disneyplus).

(Some of the topics were obviously noisier than others. Topics with the little red "n" are "noisier" than the others, meaning that a large number of tweets with a high beta in that topic were not related to the topic labels. Many tweets in the "Bad Girls Club" topic, for example, don't actually have to do with that show.)

Now remove the pieces that you highlighted. To get you started, consider:

- removing any interjections that interrupt the current topic with some extra information (e.g. , *which makes sense, because*)

- replacing contractions (e.g. *don't* and *can't*),

- reducing references to *I* to remove focus from author,

- eliminating all process language, and

- changing the beginning of any sentences that start with transition words such as *and*, *but*, or *so*

When you are finished editing the passage, note the difference in length. Is your version shorter or longer than the original? Describe the original audience and contrast it with the audience for your revised paragraph. Did the audience narrow or widen?

10 Revising: Drafts #2 Through...

Although writing a first draft can be intimidating, the revision process is often the hardest part of writing. It can be time consuming, and progress can be hard to see. This chapter provides you with a plan of attack for revising your work so that you can attain a polished draft efficiently.

Typically, we edit considerably before showing others our writing because we want a reviewer to focus on our ideas, not our writing weaknesses. However, enlisting a peer to give feedback before the final draft can give a helpful, new perspective. If we make sure to clean up the small details such as spelling, grammar, etc., a reviewer can focus on the higher-level aspects of our writing such as content and structure. The reader's role grows in the revision and editing process. Not only are we reading to take inspiration from writers we admire, now we are reading with a critical eye, looking for deviations from good practices.

Section 4.5 showed that making a plot that effectively conveyed key scientific findings is an iterative process. The iterative nature of data analysis is typically not evident in our writing. Written reports give the impression that the scientists made no mistakes in their analysis because they told us what they found and not the journey they took to get there (see Section 9.4). We saw in Chapter 6 that to storyboard we shuffled our plots and brief summaries of findings to uncover the narrative of our story. In Section 6.3, we iterated, identifying points that needed to be revisited and included in the background and discussion sections of our report.

The revision process is also an iterative process that is not reflected in the written report. It is another journey; a nonlinear complex process to refine and improve our argument. Revisions includes large-scale changes to the structure of your paper and smaller edits to sentences and paragraphs. We provide some strategies for the editing process in this chapter.

Communicating with Data: The Art of Writing for Data Science. Deborah Nolan and Sara Stoudt, Oxford University Press (2021). © Deborah Nolan and Sara Stoudt.
DOI: 10.1093/oso/9780198862741.003.0010

10.1 Preparing to Rewrite

Once we have our ideas written into a first draft, we typically refine the draft many times before it becomes a polished product. In this process, we often need to get some space from our draft. Space can be created in both time and format.

Time permitting, it can be helpful to work on something other than our paper for a few days. After we have spent some time away from our paper, we can revisit our writing with fresh eyes. Before revisiting the draft, we suggest that you write down what you expect your reader to think about the topic before and after reading your report. Then, you can reread the paper and see if your intention matches the reality.

Changing the spacing and printing the draft so that the draft looks physically different can be helpful to build distance in formatting. It can be easy to overlook mistakes on a computer screen, especially when reading the same thing over and over again. Reading out loud can be useful in this situation. When we read in our head, we often subconsciously fix mistakes without recognizing that they are there. When we read out loud slowly, we notice small grammatical details as well as higher-level problems such as ideas that do not flow or abrupt transitions. As you read aloud, we recommend that you don't stop to fix anything. Instead, mark what you want to fix with a small comment on what you have in mind and then come back to it later. You will want to examine the draft as a whole without getting sidetracked with the specifics.

Once we have caught the more obvious flaws in our writing, it can be helpful to revise with more specific goals in mind. By working with different parts of the paper in different revision stages, we also ensure that we don't fatigue of rereading the whole document over and over again.

10.2 Strategies for Targeted Revision

After proofreading the whole draft at least once, it can be helpful to target problem areas in different rounds of revision. Focused reading can help us catch issues that we may not notice when we read with a more global emphasis. Thinking about our writing weaknesses can help focus our revisions. These weaknesses may be identified from feedback we have received in the past or our impression of our writing ability so far. We can reread our draft with each of these specific purposes in mind, one at a time. The following examples give some common weaknesses and revision strategies to help detect instances of each weakness.

When you suspect that your writing style is too wordy, we suggest reading your work out loud. Note when you run out of breath. Alternatively, go through the draft looking only for unnecessary words and phrases to cut.

Example: *Wordiness*
The following block of text looks like a full paragraph, but it is actually one long sentence. It can be easy to keep adding clauses while writing because we want to make sure to say everything we know about the topic. After the initial draft, assess what is

important, delete the rest, and break the sequence of clauses into multiple sentences for easier reading.

To show that we cannot approximate our model of interest using a more flexible logistic regression, suppose we simulate data from a probability distribution in the sub-model with fixed values of the covariates and we sample with replacement from this simulated dataset multiple times to show the variability in the estimates as we might do in practice to understand the uncertainty in our estimates.

This paragraph-long sentence is unreadable! We break it up into three sentences:

To show that we cannot approximate our model of interest using a more flexible logistic regression, suppose we simulate data from a probability distribution in the sub-model. We use fixed values of the covariates and sample with replacement from this simulated dataset multiple times. These simulated datasets show the variability in the estimates.

Now it is easier for a reader to follow along. Note also that when we split the content, to maintain clarity we use *these simulated datasets* instead of an ambiguous pronoun. ■

When you suspect that your writing is too monotonous, select a few paragraphs and count the number of clauses in each sentence. Are there too many sentences in a row that have the same number of clauses? Are there too many simple sentences with just basic noun-verb pairings?

Example: *Taciturnity*

It can be easy to overcorrect lengthy sentences and break the message into too many simple sentences. Consider this revision of the paragraph from the previous example.

We cannot approximate our model of interest using a more flexible logistic regression. We show this by simulating data from a probability distribution in the sub-model. We use fixed values of the covariates. Then we sample with replacement from this simulated dataset multiple times. These simulated datasets show the variability in the estimates.

These sentences all have one clause that leads to a monotonous style. Below, we introduce a few different sentence structures for variety.

By simulating data from a probability distribution in the sub-model, we can show that even a more flexible logistic regression cannot approximate our model of interest. We use fixed values of the covariates and sample with replacement from this simulated dataset multiple times. These simulated datasets show the variability in the estimates.

Note that in the reorganization we also remove the emphasis on *we*. ■

If you suspect that your writing is disorganized, underline the main points within the article. Do they logically follow from the material presented before them? Are they hidden or in a position of prominence in the paper?

Example: *Disorganization*

In the following two paragraphs, the three main points are:

- Google uses GPS location signals and WiFi connections to tell where Google Maps users' mobile phones are, and can determine from there whether or not a user is in a store.

- Information is only collected from those who allow Google to have access to their locations all of the time, which may not be a representative sample of all people.

- Selection bias (who owns a smartphone and who uses location services on their phone) could be a big problem.

Note that the first point comes too soon; we do not have enough context to evaluate the statement. The second point is buried in the middle of a paragraph, and the third comes too late, after the discussion of its implications.

Google uses GPS location signals and WiFi connections to tell where Google Maps users' mobile phones are and can determine from there whether or not a user is in a store. You may have noticed that when you search online for a store or restaurant, Google shows you a histogram of how popular the business is at a particular time of day or day of the week. You'll even be able to see how long people typically spend there. How does Google create these histograms? For businesses that receive more traffic, Google can calculate how many people, on average, are at each business at certain times of the day for each day of the week.

If you want to find out how long you'll have to wait for a popular restaurant in an urban area, go with what Google says. But, otherwise, you'll want to take Google's suggestions with a grain of salt. Lower-income citizens and those living in rural areas are less likely to have smart devices. Information is only collected from those who allow Google to have access to their locations all of the time, which may not be a representative sample of all people. Maybe that mom and pop restaurant down the street only has a few customers who have their location services on, and these customers only come at the same time every morning. Or maybe many of the customers who go to the restaurant for the early-bird special don't have smartphones so they're not being tracked and Google underestimates how busy it is at the restaurant. Selection bias (who owns a smartphone and who uses location services on their phone) could be a big problem.

We rearrange the ideas in the following paragraphs so that the first point comes after some context, the second point begins a section, and the third point introduces its implications.

You may have noticed that when you when you search online for a store or restaurant, Google shows you a histogram of how popular the business is at a certain time of day or a certain day of the week. You'll even be able to see how long people typically spend there. How does Google create these histograms? Google uses GPS location signals and WiFi connections to tell where Google Maps users' mobile phones are, and can determine from there whether or not a user is in a store. For businesses that receive more traffic, Google can calculate how many people, on average, are at each business at various times of the day, every day of the week.

Information is only collected from those who allow Google to have access to their locations all of the time, which may not be a representative sample of all people. Maybe that mom and pop restaurant down the street only has a few customers who have their location services on, and these customers only come at the same time every morning. Or maybe many of the customers who go to the restaurant for the early-bird special don't have smartphones so they're not being tracked and Google underestimates how busy it is at the restaurant. Selection bias (who owns a smartphone and who uses location services on their phone) could be a big problem. Lower-income citizens and those living in rural areas are less likely to have smart devices. If you want to find out how long you'll have to wait for a popular restaurant in an urban area, go with what Google says. But, otherwise, you'll want to would take Google's suggestions with a grain of salt.

This rearrangement reduces the burden on the reader since it takes less work for the reader to piece the argument together. ∎

Our advice for diagnosing a lack of clarity is to identify the pronouns in your paper and assess whether they are ambiguous. When in doubt, rewrite using specific nouns.

Example: *Lack of Clarity*
The paragraph below has several instances of *it* that are ambiguous.

You may have noticed that when you search online for a store or restaurant, Google shows you a histogram of how popular the business is at a certain time of day or a certain day of the week. You'll even be able to see how long people typically spend there. If it is legitimate, it could prove invaluable to students who understand all-too-well the frustration of making the trek to a coffee shop to study, only to be unable to find a seat. But is it legitimate? First, how does it work?

We replace *it* with more informative nouns.

You may have noticed that when you search online for a store or restaurant, Google shows you a histogram of how popular the business is at a certain time of day or a certain day of the week. You'll even be able to see how long people typically spend there. If this popularity by time information is legitimate, it could prove invaluable to students who understand all-too-well the frustration of making the trek to a coffee shop to study, only to be unable to find a seat. But is the information legitimate? First, how does the histogram creation process work?

Now the reader has no doubt about the subject of each sentence. (Note that we keep *it could prove invaluable* because we have defined the noun earlier in the sentence.) ■

To diagnose abruptness, read your work out loud, and listen for what sounds awkward. Pick a few paragraphs at random in your paper and read them out of context. Are there enough contextual clues in your transition phrases to orient the reader?

Example: *Abruptness*
Below is an example of the end of one paragraph and the start of the next paragraph. There is an abrupt change in context between the two.

A similar controversy occurs in the broader debate about using species distribution models to estimate properties of interest. Given the same types of data, there are examples in the ecology literature that obtain estimates as well as counterexamples where the properties are not estimable. Arbitrating between the two conclusions is still an open problem.

In the econometrics literature, there is controversy over whether or not we can estimate parameters accurately in the presence of non-random missing data. ...

We add a transition sentence to the second paragraph that reiterates the conclusion of the previous paragraph and introduces it in another context. Note that we do not start with *A similar controversy* again because we already used that transition phrase earlier.

Just as the richness of certain types of data is open to debate in the ecology literature, data quality is also contestable in other fields. In the econometrics literature, there is controversy over whether or not we can estimate parameters accurately in the presence of non-random missing data.

Connections between ideas may seem obvious to us because we have thought deeply about our work. However, a reader may be less familiar with the content so we explicitly guide them through our thought process. ■

If you suspect that your writing lacks focus, read through your article looking for backtracking, i.e., places where you return to something you already talked about. Note all the places where the subject changes, including places where you prematurely jump ahead. Do you ever start one idea, jump to another one, and then come back to the original idea? A lack of focus can easily happen in early drafts when our main goal is to get all of our ideas written down. In the revision process we can see the whole picture and reorganize to avoid this stream-of-consciousness style.

Example: *Lack of Focus*

The following paragraph bounces back and forth between ideas rather than organizing the ideas and presenting them in a logical, compelling order.

A mistake I notice is that my students are not good with word problems. It's pretty hard to visualize and understand information from word problems. For example, if a question used the phrase "How much more," you may not know that you will compare two things, probably using subtraction. Since some kids don't know what the phrase means in context, I would help them draw and ask guiding questions to take it step by step. There are other phrases in word problems too, but teaching them how to visualize by drawing out the pictures helps a lot. Another tricky type of word problem is converting between units, like 12 inches is equal to 1 foot. Students do not know how to visualize length and do not understand how units work. It's good to have examples to show them the difference between the units, like a ruler or measuring cup.

Here we group ideas and order them so that the narrative moves in one direction.

I notice that word problems are tricky for my students. I think this is because they have a hard time visualizing the problem, interpreting certain phrases in the problems, and avoiding pitfalls such as changes in units. For example, if a question used the phrase "How much more," students may not understand that this is a comparison problem involving subtraction. Similarly if the problem involves a conversion between units, like inches to feet, students may not be able to identify the main objective of the problem. Both problems can be made more concrete with examples. Drawing a picture of the groups being compared can help show the difference in size in terms of "how much more." Physical objects, such as a ruler or measuring cup, can help show the difference between units. My strategy to help students is to have them draw the problem, ask guiding questions, and then go step by step.

This example shows how complex a revision can be. ■

10.3 Editing with Others

After a few rounds of our own proofreading, editing, and revising, it can be helpful to get a fresh set of eyes to look over our paper. A peer represents a reader. They can identify gaps that go unrecognized given our immersion in the project. Peer review is a two-way street. Offer to read something of your peer's writing in return and give the type of feedback you would like to receive.

10.3.1 Giving Feedback

When asked to give feedback to a peer, we recommend asking them some questions to ensure that the review will be focused and helpful. Knowing where they aim to submit their work can help assess whether the structure is appropriate and if the content is presented at the right level for the intended audience. It can also be helpful to know what aspect of the paper they are most concerned about. This will help target the review and focus our energy on the weaker aspects of the draft.

Before reading the draft, it is helpful to make a plan of what to look for. If the review is for a journal instead of a friend, review the journal guidelines for how to evaluate the work. When reading something that may be outside of our area of expertise, a plan can help us review more efficiently.

As you read a peer's work, consider the following prompts to guide your review.

- First, skim through the paper to get a sense for the organizational structure. Are all the major components of the report (see Chapter 7) accounted for?

- Next, read through the paper without making any edits. Summarize the paper in a few sentences.

- What are you confused about and why? (i.e., what remains unclear?)

- What are you curious about and why? (i.e., what details would you like to see added?)

- Read through the paper again, and now make comments and edits. Underline sections that are particularly clear/well written, and try to articulate why.

- Try to avoid copy-editing. Focus on organization and flow rather than comma usage and spelling.

- Look through your comments and edits. Organize your feedback into themes. Try to avoid just giving an unstructured list of comments. What is one thing you would have done differently and why?

- Don't be afraid to give serious constructive feedback. You aren't being mean; you are being helpful.

- End your feedback by identifying the biggest strength of the paper.

Going through the review process for someone else can help you both better proofread your own work and understand the feedback you receive from others.

10.3.2 Receiving Feedback

Receiving feedback can be difficult. However, we should remember that feedback is meant to help us strengthen our writing, not act as a judgement of our writing ability. Therefore, we should endeavor to accept feedback with grace and not take any comments personally.

Since we are naturally invested in our work, we can be resistant to making the major changes that a reviewer advocates. We can first read through the comments, and take a minute to be defensive. What are the reasons for ignoring the advice of the reviewer? Then we can take a higher-level approach and consider where the reviewer is coming from. Finally, we identify the common themes within the feedback. Are the main criticisms about content, big-picture-writing aspects such as structure and clarity, or detailed aspects of the writing such as word choice or sentence flow?

Creating a revision to-do list that is specific and contains all of the comments made by our reviewer will help us stay organized. As we make changes, we keep track of what we do to address each element of feedback (see Portfolio Piece K) to convince our reviewers that we took their comments seriously. Below, we provide examples of reviewer comments, categorize them, and make a plan for addressing them.

Example: *Details*
Consider the following two reviewer comments; both concern details.

This manuscript would benefit from closer proofreading. For example, what do you mean by "As we collected data, we periodically discarded entries that did not have location data and contain one of our attributes of interest." Does "not" modify both clauses? Another example is the extra "given" in "What are the summary statistics of the given average temperatures given?"

Additional documentation for the supplementary code is needed. A few errors were thrown when I tried to run the code.

To correct proof-reading issues, we can print our article and read it out loud, as described in Section 10.1. We also can have a peer do the same to catch unclear phrases and typos.

If we are surprised by the coding errors the reviewer ran into, we should double-check dependencies that we may have failed to disclose. We can use someone else's computer to test our code. After we have figured out the difficulties that the reviewer had in running our code and fixed any errors, we should update the documentation to clarify, e.g., describe the expected inputs and outputs for each function. (See Section 5.2.2.)

Both of these reviewer comments tell us that we should be more careful. We need to improve our proofreading and better test our code. ∎

Example: *Narrative*
The comments below pertain to the emphasis and narrative of our article. The first concerns the visualizations.

Given that a key aspect of the paper is data visualization, one way to improve the paper would be to suggest a few more graphical displays. More specifically, images that are more exciting and tell a stronger story about the data would be useful.

The second comment indicates that the reviewer is not satisfied with the data description.

It seems to me that the main uses of this dataset are to illustrate the many difficulties involved in collecting good data. A few of these difficulties are discussed briefly, but I would advocate an expansion of these sections.

We could strengthen our argument by focusing on the challenges inherent in the type of data we are working with and making our graphics more compelling. To do this, we might revisit the figures in the paper and determine whether the story is clear in the visualization (using strategies from Section 6.1) and whether the captions are as informative as possible (see Section 7.3.5). We should also consider whether a different type of plot might better emphasize the story (see Chapter 4). After editing the figures, we can give them to a peer to see if they understand what the graphs are trying to say and if they find them compelling.

As for the data description, we can start by brainstorming difficulties in the collection process. We might draft a new section that focuses on these ideas. If we add a new section, then we need to consider where this section best fits in the paper and adjust the transitions as needed. ∎

Example: *Content*
The comment below addresses content. According to the reviewer, we have not emphasized the limitations of our data enough. In response, we should aim to be more precise.

The authors warn that the data are not a random sample of the U.S. population, but there are additional limitations for inference based on the data collection that should be addressed. I recommend a stronger emphasis on using this data for exploration rather than inference.

We should read through the paper looking for places where language could be interpreted as inferential when this is unwarranted. We can adjust language so exploration is the focus. In the discussion, we can add a paragraph that points out the limitations of the data and explicitly re-emphasizes that the data is meant for exploration purposes rather than inference. ∎

In some cases, a reader's suggestion may not fit well for our paper. Then it is important to be able to justify our choice to keep the original writing. For example, the feedback may be based on a misunderstanding about an aspect of our work. If we spend some time rewriting, then we may be able to address the feedback by clarifying the confusion rather than making the suggested change explicitly.

Example: *When the Reviewer is Wrong*
Here is an example of a reviewer comment that we disagree with.

> In your paper you state that "Another author proposes a solution that disentangles the parameters of interest using the limited data." but then you later contradict yourself by asserting that "the limited data is not fit for our purpose".

We acknowledge that this confusion arose because we were unclear in our original draft. We clarify what we mean and then explain why the reviewer's suggestion is not appropriate.

> The contradiction is due to the lack of clarity in our original statement. The approach in the literature allows parameters of the model to be estimated, but the properties of the underlying data generating process cannot be estimated. Therefore we conclude that the limited data is not fit for our purpose. We have changed the first quoted sentence to be more clear: Another author proposes a solution that is able to estimate the parameters of the model using the limited data, but properties of the underlying data generating process are not able to be accurately estimated. ■

Sometimes, our best peer reviewers are those outside of our area of expertise because they help us see how our paper would appear to a less familiar audience. However, there are some more technical aspects of a paper that another scientist in our field is best suited to review. The next section discusses the aspects of a draft that a fellow scientist is most equipped to review.

10.4 Revising the Argument

In the revision process, we should consider whether we have made a convincing and compelling argument to the reader about our findings. To help us make this high-level revision, we rely on the lessons learned from the previous chapters of this book. Below are questions designed to help you address this broader goal.

Context Have you provided enough context for the intended audience as discussed in Section 7.5.1? Will the reader understand what has been done before on your topic and what gaps still remain in the literature? Will the reader have a sense of where your work fits into the literature and/or fill any gaps?

Broader Impact Are the broader implications adequately described (Section 7.5.2)? Will the reader understand your motivation? Will the reader know what impact your results have in context as well as methodologically?

Transparency Are the methods transparent and replicable as discussed in Section 5.5? Are your data processing and analysis choices throughly explained and defended? Would the reader be able to reproduce your approach with guidance from both your report and any supplementary materials?

Visual Evidence Do your visualizations reveal the key insights of your findings? Do the figures follow the best practices provided in Chapter 4? Do the captions meet the criteria discussed in Section 7.3.5?

Interpretation Are the results interpreted fairly or are they overstated (Section 7.3.4)? Does the description of your results focus on the aspects that are most important to your question or problem of interest? Do you avoid words and phrases that imply more certainty than you have?

Generalization Are the limitations of the methodology and the generalizability adequately described as discussed in Section 7.4.1? Will the reader be aware of the problems you ran into? Will a reader unambiguously know where the results can and cannot be extended?

These questions may lead you to consider a significant overhaul of your report. Don't shy away from this task. Reorganizing and rewriting often takes less time than imagined and it can lead to dramatic improvements in your writing. With practice, you will need to make large changes less frequently (even experienced writers make major overhauls of their work), and you will write both more efficiently and effectively.

10.5 Revising for the Intended Audience

When we revise, we also want to keep in mind the journals that we are interested in submitting our work to (Section 7.2). Here are some things that editors and reviewers may look for.

- Is the main point revealed early on? Editors often skim the beginning and make a decision about whether your paper will go further in the review process. Make sure they can see the point of the paper early on.

- How is this work different from others'? If your work appears to be too similar to other papers, editors might not see the reason to publish your work. This is why it is important to explicitly show what others have done, identify the gap in the literature, and make clear why your work fills the hole.

- Does the work matter? If your work lacks an answer to "so what?", a reviewer may not understand the full merit of your findings. Your reviewers will most likely be outside of your main specialty. Be sure to connect your work to something others can appreciate even if they are not in your specific sub-field.

- Is the argument persuasive? If there are gaps in your logic, reviewers will identify them and cite them as reasons to reject your paper. Make sure you have anticipated possible counterarguments and addressed them proactively.

- Is the paper concise? Page limits matter (as well as formatting guidelines). Journals receive many potential articles, and it can be easier to first filter based on who follows the directions in the author guidelines. You do not want to have a paper rejected without anyone even reading it, so spend time pruning your paper if it exceeds the length limit.

- Is the paper written at the appropriate level for the intended audience? Consider the audience of the journal and make sure your language choices match the experience of the readers. Are you using jargon without defining it? If reviewers are confused, your message will be diluted.

- Does the intended audience match the readership of the prospective journal? Even if your paper passes every other test, if the content is not presented in a way that is accessible for the audience of the journal, the paper will not be accepted. Spend some time looking through the overview of the journal's aims and skimming other articles in a recent edition of the journal.

Editing by ourselves, with others, and with an eye towards what scientists in our field look for in a good article can help bring a first draft to a version that is ready for publication. The activities at the end of the chapter will help you practice.

10.6 Notes

Tips about revising your writing can come from a variety of sources including Alley's *The Craft of Scientific Writing* (e.g., needing "distance" from a draft) [1]. Although Scofield [5] talks about revision for novels, many of the recommended techniques can be applied to technical writing. The idea that writing weaknesses pertaining to the details and mechanics of writing distracts from the effectiveness of writing was inspired by Chapter 5 of Bean's book *Engaging Ideas: The Professor's Guide to Integrating Writing, Critical Thinking, and Active Learning in the Classroom* [2]. The idea of thinking about what your reader will think about the topic before and after reading your work comes from Chapter 3 of Bean's book [2].

Examples throughout this chapter were adapted from student work (including blog posts by Phu [4] and Yang [8]) and our own. The sample review comments in Section 10.3.2 are adapted from reviewer comments Sara has received in the past.

The questions to answer for peer review in Sections 10.3 and 10.8.2 were partially inspired by a response template for peer review shared with the authors by Kathleen Donegan (UC Berkeley English Department), teaching resources from University of Michigan Sweetland Center for Writing [7], and sample peer-review materials from

Brandeis University Writing Program [3]. The idea for a writing journal in Section 10.8.3 was inspired by Chapter 3 of Silvia's book *How to Write a Lot* [6].

For more practice, consider working on Portfolio Pieces D or K.

10.7 References

[1] M. Alley. *The Craft of Scientific Writing*. Springer-Verlag, New York, 3rd edition, 1996.

[2] J. C. Bean. *Engaging Ideas: The Professor's Guide to Integrating Writing, Critical Thinking, and Active Learning in the Classroom*. John Wiley & Sons, Inc., Hoboken, NJ, 2011.

[3] Brandeis University Writing Program. Sample peer review sheets. https://www.brandeis.edu/writingprogram/teachinguws/instructorresources/peer.html.

[4] B. Phu. Mistakes that kids make when they do math. https://stat198-spring18.github.io/blog/2018/04/17/math-mistakes, April 2018.

[5] S. Scofield. *The Last Draft: A Novelist's Guide to Revision*. Penguin Books, London, 2017.

[6] P. J. Silvia. *How to Write a Lot*. APA LifeTools, 2nd edition, 2018.

[7] University of Michigan Sweetland Center for Writing. Using peer review to improve student writing. https://lsa.umich.edu/sweetland/instructors/teaching-resources/using-peer-review-to-improve-student-writing.html.

[8] S. Yang. How does Google's popular times work? https://stat198-spring18.github.io/blog/2018/04/25/how-does-google-popular-times-work, April 2018.

10.8 Activities

10.8.1 Focused Revision at Sentence Level

The revision process can be a daunting task. Focusing on certain aspects of writing, one at a time, can make revision more manageable. By identifying a set of qualities that we want our writing to have and going through our work in a targeted way, we can catch more issues and more effectively revise. Similarly, if we know where our weaknesses lie, we can be extra vigilant when editing.

Identify three qualities you admire in others' writing. Provide three examples of each quality. Then try to find at least one example of each quality in your own paper draft. Assess your writing weaknesses. Which of the categories from the examples in Section 10.2 most resembles your writing? Pick at least two targeted revision strategies that are recommended for your weaknesses and revise your paper with them in mind.

10.8.2 Formal Peer Review

Exchange a paper draft with a peer. For this activity, you will practice both reviewing another's paper and writing a formal review as you might do as a reviewer for a journal.

1. Write an introductory paragraph that summarizes the content of the paper.

2. Move on to the strengths of the piece.

   - What do you find compelling?
   - What is especially clear?

3. Then move to the weaknesses.

   - What is missing?
   - What could be made more clear?
   - Are there any inconsistencies?

4. Wrap up your review with some big-picture comments (e.g., about the overall structure, effectiveness of argument and visualizations, word choice, etc.) and emphasis on the few key things that most need to be changed or strengthened.

Your review should be about one page, single-spaced, not including the list of typos and grammatical errors.

Other things that you may want to consider are:

- overall structure,
- transitions between paragraphs and ideas,
- word choice,
- effectiveness in communicating to the given audience,
- appropriateness of the methods used if applicable, and
- effectiveness of visualizations if applicable.

10.8.3 Writing Journal

When we are in the revision stage it can often feel like we aren't making any progress. This activity aims to help you understand how you use your time and keep track of your accomplishments throughout the revision process. By tracking your writing goals and productivity for a week, you will better appreciate your efforts and learn how to use your time more efficiently.

1. First, track your schedule.

   - Write down when you have consistent pockets of time throughout the week.
   - Block out a total of six hours in your schedule to write. You can stop after four hours of seriously productive time.

2. Next, make concrete goals for your writing time.

- In each of your blocks write down concrete goals. Goals can include:
  - Make an overall outline.
  - Develop a piece of an outline into a more detailed outline.
  - Write a particular section.
  - Revise a certain section.
  - Revise as a whole.

3. As you work, track your productivity. (Note: this is not a time to be writing code.)

- Keep track of what you get done in each block of time in a spreadsheet like the sample in Table 10.1.
- Note the time when you start and your goal for the time period. Do a word count before you start.
- Every half-hour describe briefly what you accomplished in that time frame. Do a word count.
- When you end your block of time, do a word count and write whether or not you met your goal.
- If you are not able to make progress during a writing-time block, it is fine to write "I didn't get anything done".

4. After you write, reflect on how your writing time went.

- For the times that you didn't get anything done.
  - What were the barriers to writing?
  - Were there distractions?
  - Were you feeling "stuck"?
- What did and did not work about your writing schedule?
- How would you change it to try to be more productive?

Date	Start Time	End Time	WC Start	WC End	Goal	Accomplished
11/27	11:00	11:30	0	220	Outline report	All but methods and results sections
11/27	12:30	13:00	220	450	Outline report	Methods and results sections
11/28	9:00	9:30	0	250	Write methods and results sections	Got half-way through
11/28	9:30	10:00	250	520	Write methods and results sections	Got half-way through
11/28	10:00	10:30	0	0	Write introduction and literature review	Didn't get anywhere
11/28	10:30	11:00	-	-	-	Skipping because I hit a wall
11/29	13:00	13:30	0	230	Write introduction and literature review	Got half-way through
11/29	13:30	14:00	230	350	Write introduction and literature review	Finished
11/29	14:00	14:30	520	475	Edit so far	Edited methods and results sections
11/30	15:00	15:30	350	300	Edit so far	Finished what started on 11/29
11/30	15:30	16:00	-	-	-	Skipping because I reached my goal, hooray!
11/30	16:00	16:30	-	-	-	Skipping because I reached my goal, hooray!

Table 10.1 Example Writing Schedule.

Part V
Science Writing and You

L earning how to prepare both a formal and informal piece of writing about your data-analysis work is just the start of communicating. This part of the book focuses on communication beyond the page and gives advice about wielding your identity as a science writer.

In Chapter 11, we discuss alternative venues for communication, including oral presentations and online communities. We make connections between the strategies for writing that we have discussed previously and explain how they can be adapted to help you prepare to communicate your work in person and online. In particular, we show how powerful and versatile the storyboarding approach can be.

We also provide strategies for fostering both professional and personal networks, which can help you communicate to a wider audience. Just as your writing voice develops with continued practice, your identity as a scientist develops as you gain more experience. This chapter shows how to weave your professional brand together with your personal identity to further develop as a scientist.

In Chapter 12, we provide additional opportunities to practice your writing. More extensive prompts are included to help you further practice your writing skills and build a portfolio of writing samples.

11 Embracing Your Role as a Scientist

Polished draft in hand, you may ask yourself, now what? It is now time to (finally) share your work with the world. So far, this book has aimed to teach you how to read as a writer, prepare to write as a writer, write as a writer, and edit as a writer. We hope you now feel comfortable identifying as a writer (because you definitely are one). Now we want to give some advice about wielding the identity of a writer as you advocate for your science.

Despite being both a scientist and a person existing beyond our work, we often must separate our personal feelings from the work we do to remain objective. It can be helpful to have a network (a source of professional support), a research focus (professional interests), a community (a source of personal support), and an identity (personal interests) to help balance between the professional and personal. Your community helps support your communication within your own circle of influence while your network can help you reach across community boundaries. Your research focus and identity help you navigate the social aspect of research by signaling what interests you.

This chapter discusses opportunities to embrace the social aspects of communicating your work and gives advice about venues for sharing your work beyond the page.

11.1 Expanding Our Professional Network

To expand our professional network we often must physically leave our place of work or study to meet others and showcase our work. These meetings may occur in annual conferences for professional societies or more informally at "meet up" groups (i.e., in-person events organized online by people with similar interests).

When we seek others to add to our network we can't expect someone we have just met to read our paper while we stand there waiting for them to get caught up on what we are working on. Instead, to take advantage of the meeting we rely on oral communication

Communicating with Data: The Art of Writing for Data Science. Deborah Nolan and Sara Stoudt, Oxford University Press (2021). © Deborah Nolan and Sara Stoudt.
DOI: 10.1093/oso/9780198862741.003.0011

to give them an overview of what we are working on. This type of communication can take the form of a formal talk, a lightning or speed talk, or a poster. It can also take the form of a one-on-one conversation.

Up until this point, we have talked about written communication, but many of the strategies and advice we gave for preparing written material applies to preparing formal talks and being ready to chat informally with new people. Each of the formal oral communication venues mentioned above has an analogous written form. A formal talk is like a formal paper, a lightning or speed talk is like a press release, and a poster is like a blog post. More informal networking opportunities happen in unstructured social settings and don't have an obvious writing analogue.

11.1.1 Formal Talk

An effective talk is like an effective paper; it has an organized structure, a well-defined audience, a compelling narrative, and is delivered clearly. A talk has many of the same elements as a paper (review Section 7). We start with an introduction of ourselves and the background of the problem we are working on, we give some grounding in the literature (e.g., what has already been done, what holes are we trying to fill), we launch into the heart of our solution/approach, and we conclude with some discussion of the impact of our approach and an announcement of any future work we have planned. Just as we would storyboard before drafting a paper, we can also storyboard before a talk (review Section 6). In some ways, the storyboard is even more natural in the case of a talk that involves slides. Each note card represents a slide, and we can physically arrange them to design the flow of the talk.

The level of detail provided in each part of the talk and the relative length of each part depend on the venue (review Section 7.2 for a discussion of audience). If we are at a conference for our particular sub-field or in a section dedicated to our sub-field, we may assume our audience will have a lot of the background needed to follow along with our work. We can then focus more of our attention on our approach. At a conference with a broad theme or in a more interdisciplinary venue, we may want to heavily emphasize the background and context while only providing the high-level version of our approach.

Just as we follow the journal length requirements for a formal piece of writing, in a formal talk we conform to the venue time requirements. Having the right talk length is essential. We do not want to be rushing through slides because we have too much material to cover, and we don't want to run out of things to say. A good pacing guideline is to have no more than one slide per minute of talk time allotted. You may spend an extra minute or two on a few key slides. Once our slides have solidified in order and content, we recommend that you practice the talk and time yourself. Make sure you have time left for questions (a good rule of thumb is about five or ten minutes of questions in an hour-long talk). Also consider that you might talk a little faster during the actual talk if you are nervous; remember to talk slowly, even if it feels too slow.

One important distinction between a paper and a talk is that we do not want us or our audience to be doing a lot of reading during our talk. Keeping our slides uncluttered with minimal words and one or two figures will ensure the focus is on what we are saying. Including visualizations, rather than blocks of text, can help guide a listener through our narrative and keep their attention. Just as we don't want to give our listener blocks of formal text, we do not recommend that you write out the talk and read it word for word. Having an outline and perhaps a few notecards with phrases to jog our memory if we get stuck are fine, but relying on rigid text in a talk format makes it more obvious if we slip up. We want to aim for a natural delivery, as if we were just talking with a friend. This ease comes from practice and preparation, not memorization.

Remember that like our paper, the talk is a showcase for our work. The important thing is that after the talk, listeners can follow-up with us to ask questions and find out more about our work. We recommend that you make follow-up easy for the audience. Provide contact information on your slides, and if possible, remain at the venue location after your talk to mingle and give people the opportunity to ask questions one-on-one rather than in front of a crowd.

11.1.2 Lightning/Speed Talk

A formal talk showcases our work in a venue where there is ample time to address at least some details. In contrast, a lightning or speed talk (often only about five minutes long) is an abbreviated advertisement for our work. We cannot possibly explain our entire project in 5 or 10 minutes, so we need to give the audience enough detail to understand the big picture and be invested in the answer to our question. We recommend that you think of this style of talk as an elevator pitch for your work; make the listener want to come and talk to you afterwards to learn more.

We have a short amount of time to make people interested in talking with us further. Like a press release (review Section 2.1), this format of talk requires that we put the important information first, avoid getting into technical details, and have an enticing ending that will make listeners want to follow-up with us.

Preparation for a short talk is much like preparation for a long talk, although we do not spend any time on details. We can storyboard and practice the talk with a timer. However, it is important to note that just because the talk is short, this doesn't mean the talk is easy to give. It can actually be much harder to give a short talk without preparation than an unpracticed long talk. In a short presentation, your pace must be precise; there is not a lot of leeway for stumbling due to nerves. We recommend that you practice, practice, practice.

11.1.3 Poster

A poster is a low-stakes way to advertise our work. Sometimes, a lightning talk is paired with a poster so that listeners who were intrigued by our talk can follow-up with us. Other times, posters are there for conference attendees to browse on their own time.

Think of a poster like a blog post. A viewer stumbles upon it and must decide whether to commit to engaging or rather move on to one of the many other posts out there. To capture and hold the attention of a viewer, our poster should have a succinct story and a striking visual. Bullet points and whitespace are powerful tools. Just as we avoided slides crammed with text, our poster should also be free of dense blocks of text.

Storyboarding can also be used to prepare our poster. In this scenario, each notecard in our storyboard is a "block" of our poster. Using these blocks of content, we can explore potential road maps of how a viewer will step through our poster. Left to right? Top to bottom? A mix, separated by columns? You can decide what visually makes sense.

A poster still requires talking to people. It can be helpful to also have a few minutes of commentary planned for a viewer who prefers to be guided through the poster. We suggest walking them through your main points, answering any specific questions they have, and taking note of any suggestions that they give you.

11.1.4 Networking

When we hear "networking" we often think about social interaction with strangers in artificial settings. However, networking does not have to have a superficial connotation. Any time we talk to a person, we are networking. Writing for broader audiences should strengthen our ability to talk casually about what we do.

Conferences are not just about giving and listening to talks; networking is also a major part. Take advantage of having many people with similar interests all together in the same area. Before you go to a conference, do some preparation. Reach out to people whose work you admire and see if they would like to meet during the conference. Professional relationships can also be formed serendipitously in between sessions and during meals and coffee breaks, so push yourself to start conversations with other conference attendees.

As we chat with others, we may want to formalize a connection by referencing a product of our work. However, our work is not always in a polished state when we find ourselves in these networking situations. Instead, we can post a draft of our formal paper on the arXiv, a freely accessed archive for scholarly articles, to document and provide access to our works in progress. If we are not even at the formal drafting phase of research, we may consider writing occasional blog posts to keep track of our research process publicly.

11.2 Building a Research Focus

We have talked about expanding our professional network in the context of showcasing our work at different, in-person, venues. Connections can be made by attending conferences, visiting different campuses or departments, and taking advantage of the connections of those we already have a connection with within our network. By connecting with others with similar interests, we can find out about papers to read, job or internship opportunities to apply for, conferences to attend, and potential collaborators to work with. However, connections can also be made remotely. Academics have become active on social media (e.g., Twitter), taking advantage of its networking and information-sharing power. Online connections can be made and leveraged at a faster pace. All it takes is one "influencer" to promote us to make other people aware of our work.

In order to fully take advantage of online community connections, we must make others aware of what we are interested in and what our area of expertise is. This way they can point us to relevant materials and know who to come to for questions on a particular topic. We build a professional identity by repeatedly showcasing our work, skills, and strengths. To help carve out our niche within the broader research community, we consider our expertise. Is our research focus deep; have we invented a particular method and know it inside and out? Is our focus more broad; do we solve problems in a wide range of scenarios? Is our focus domain specific; do we feel comfortable being the go-between for quantitative and non-quantitative people? Is our focus on tools; do we build software to simplify analyses or data wrangling? Is our focus on teaching; do we explain concepts well and advocate for our students?

Whereas at individual speaking events, we focus on exhibiting one piece of work, it is also important to have a complete body of work in an easily accessed place to help people explore at their leisure. Building a portfolio of writings and code (see Chapter 12 for content ideas) and having it in an easy-to-link-to place (e.g., GitHub, your personal website) so that we can provide links to resources when asked, is a great way to display our professional interests. A full portfolio can also display how our interests and expertise have changed over time.

We cannot always travel to talk about our work with others. Using social media to alert others when we have added something to our portfolio, whether it's a talk, a blog post, or a new draft, is a less resource-intensive way to get our work noticed. We can also use social media to solicit feedback on a draft of our work or advice on a question we are pondering. By crowdsourcing the feedback process, we can get faster feedback more frequently.

Promoting our work may feel like bragging, but exposure is an important part of being a researcher. We want to be part of scientific discourse and tell others what we have been working on so that others can learn from it, give feedback, and extend it. Small announcements via social media maintain a living curriculum vitae and also gives us a sense of milestones reached. The data analysis and writing processes can be intense, so it is nice to celebrate the little victories in a less formal venue. However, if we are going to take advantage of the speed of social media to promote work and get feedback, we need to be sure to return the favor. We should support others' work and give our feedback when requested.

11.3 Fostering a Personal Community

A professional network can help us build a career, but a personal community is also a key to our success. We cannot always be in work mode, so we also need a source of support that is separate from our job.

Our personal communities can be related to our work (e.g., coding communities, question-and-answer communities) or they could be a book club, a knitting circle, a basketball team, or an improv group. We will primarily talk about communities that can be connected to your work in this section, but alternative communities are also an important source of support.

Coding communities ranging from open-source projects to Stack Overflow can both provide help for work-related efforts and support for our more creative endeavors. They can also be a way for us to give back or pay it forward by providing an opportunity for us to use our expertise to help others. "Side projects" where we use our work-related skills in a non-work-related setting can be a rewarding way to re-charge after a challenging day of data analysis or writing. The following examples provide insight into both contributing to and benefiting from these communities.

Example: *Community Standards: Codes of Conduct*
Contributions to online communities do not have to be code. We can contribute content to Wikipedia (see Sections 8.10.9, 9.8.4, and 11.7.3). However, we must recognize that Wikipedia is an online community that has community standards and protocols that we must conform to. In this excerpt from Wikipedia's guidelines for contributing, we are told about how we should conduct ourselves when we disagree with another editor.

While discussing matters, it is very important that you conduct yourself with civility and assume good faith on the part of others. Edit warring (repeatedly overriding or reimplementing contributions) is highly discouraged. There is a bright-line rule called the three-revert rule, the violation of which may lead someone to be blocked from editing to prevent further disruption. Disruptive editing is not always intentional, as new editors may simply not understand the ins and outs of Wikipedia. ∎

Example: *Community Standards: Minimal Reproducible Example*
On Stack Overflow (and similar sites), there are guidelines for how to ask a question such that you are most likely to get a useful answer. The idea of a minimal, reproducible example (reprex) is outlined in this excerpt of the guidelines.

Your code examples should be...
...Minimal—Use as little code as possible that still produces the same problem
...Complete—Provide all parts someone else needs to reproduce your problem in the question itself
...Reproducible—Test the code you're about to provide to make sure it reproduces the problem ∎

Other forms of personal community that can relate to our work are writer and accountability groups. Writer groups often are formalized, peer-review sessions (refer to Section 10.3). Writers each bring an excerpt of something they are working on and perhaps a specific aspect of the writing that they are most concerned about to the meeting and then ask for feedback from the group. You can start your own writer's group with people from your professional network. You may even want to meet virtually rather than in person.

Accountability groups provide a venue to set and report on goals to peers rather than supervisors, creating a judgment-free accountability mechanism. These goals could be a mix of work and personal. In these groups, members take turns reporting on what they accomplished since the previous meeting, reflecting on why they did or did not meet their goals, and making revised goals for the upcoming time period. Group members can also provide advice for how to overcome obstacles. It can help if the membership of this group is not made up of individuals in our inner circle since maximum benefit comes from being able to be honest about our struggles and to fail without worry of judgement.

11.4 Welcoming Who We Are

We carry our identity with us throughout all of our professional endeavors. Sometimes, our identity and our profession specifically intertwine. There are identity-based conferences (e.g., Women in Data Science, Society for Advancement of Chicanos/Hispanics and Native Americans in Science, Grace Hopper Celebration) and identity-based coding groups (e.g., Girls Who Code, Black in AI) that provide an opportunity for sharing work, advice, and support among people with a shared-identity experience. Other times, we must carve out our own place, perhaps by leaning on our personal community.

Having a visible presence, either online or off-line, regardless of the main focus of the content, can increase the visibility of traditionally underrepresented groups in a particular field. Sharing our work and experience balancing personal and professional identities can bolster others who may be interested in a similar field but do not see many like them in it.

Example: *Blogging to Bring Perspective*
In this blog post excerpt, Daly tells her personal story of her first time at a conference. She does not specifically reference identity, but she is honest about feeling anxious and overwhelmed, helping to normalize those feelings.

This was my first time at the annual rOpenSci Unconference event held in Los Angeles, CA and incidentally my first time in the city. I was excited at the onset, from getting the invite email, to closely watching the project proposals pitched during the lead up to the event. I had felt slightly anxious to meet so many useRs that I had only really engaged with on Twitter, but a far cry from the how nervous I felt when attending the international useR conference in Stanford, CA last Summer. My nerves were calmed after the interesting icebreaker on the first day where I got to group up with other attendees based on animal preference (#rdogladies & #Rpuppies obviously!), learning styles, and thoughts on my personal role within R community (tl;dr: I'm one of many folks suffering from imposter syndrome regardless of my participation as Co-organizer of R-Ladies CT, and Core Member & Survey Lead of Forwards).

In the beginning, I did not feel as productive and was really bogged down with the notion of other groups bountiful commits and producing a full-fledged R package in less than 20 hours. I got some sage advice form the community manager, Stefanie Butland about taking mental breaks and developing a smaller project scope. I also appreciated the breaks to de-stress from the social interaction during dinner on the first night which enabled me to achieve some progress on my project in a secluded setting that I was familiar with.

This blog post acknowledges networking nerves. Readers can benefit from this vulnerability and moderate their own feelings of nervousness around new people. ∎

Example: Blogging to Share Background
In this blog post, D'az shares his experience as a first-generation tech worker with the explicit goal of helping others facing a similar situation.

I have seen many under-served tech workers talk about struggling with working too much, feeling guilt, or a sense of paranoia when it comes to their newfound wealth. This is especially hard if you are the first generation to have any kind of financial freedom. I have had trouble making this adjustment from poor to rich (but not wealthy), and I hope that sharing my adjustments helps others in similar situations.

This blog post teaches us about the complexities of navigating between past and current identities. ∎

Curating our professional persona is useful for our career, but it is also necessary to reflect on our personal values and goals. What drives us to do the work that we do and what matters to us beyond our work? As we communicate our work, we should take note of what energizes us and follow that energy as we move forward in our careers.

Although we can give you tools and advice for *how* to communicate your work effectively to a variety of audiences, we cannot guide you in *what* you ultimately say. We hope that by embracing your dual role as scientist and writer and using the skills you have developed by working through this book, you will go on to advocate for what you are passionate about in both your professional and personal spheres.

11.5 Notes

Conferences often have a set of guidelines or advice for talks and posters. For example, see the American Statistical Association's advice for speaking [2] and presenting a poster [1] at the Joint Statistical Meetings. The Software Sustainability Institute provides specific guidance for lightning talks [24]. *Science* magazine provides many resources about networking [18], and *Nature* has advice especially for introverts [10]. The arXiv [3] is an open-access repository for scientists to make their scholarly articles publicly available.

The Chronicle of Higher Education has written about the tie between social media and scholarship in a variety of articles including by Carrigan [6], by Mangan [15], and by Martinez [16].

Coding communities have many resources online. For more examples of codes of conduct, see rOpenSci [21] and the NumPy Community [23]. For more guidance in creating minimal, reproducible examples, consult the reprex package in R [5] and the reprexpy project in Python [4]. The Wikipedia code of conduct can be found at [35], and the community standards for Stack Overflow are available at [25].

Daly describes her experiences at the rOpenSci Unconference in her post [7], and D'az shares his experience as a first-generation tech worker in his post [8].

Johns Hopkins Medicine provides many resources about writing accountability groups [14] for faculty that can be repurposed for students, and Rockquemore [20] discusses a variety of styles of writing groups.

A list of national organizations for diversity in science, technology, engineering, and Mathematics (STEM) was compiled by the Icahn School of Medicine at Mount Sinai [12].

The sample biographies in Section 11.7.1 were written by our students [26].

For more practice, consider working on Portfolio Piece V.

11.6 References

[1] American Statistical Association. Poster presentation tips. https://ww2.amstat.org/meetings/jsm/2020/postertips.cfm, 2020.

[2] American Statistical Association. Presentation tips. https://ww2.amstat.org/meetings/jsm/2020/presentationtips.cfm, 2020.

[3] arXiv. About arXiv. https://arxiv.org/.

[4] C. Baker. reprexpy. https://pypi.org/project/reprexpy/, January 2019.

[5] J. Bryan. How to use reprex. https://reprex.tidyverse.org/articles/articles/learn-reprex.html, September 2018.

[6] M. Carrigan. Social media is scholarship. *The Chronicle of Higher Education*, October 2017. https://www.chronicle.com/article/Social-Media-Is-Scholarship/241467.

[7] J. Daly. My Experience at the 2017 rOpenSci Unconference. https://jasminedaly.com//2017-05-28-runconf17-experience/, May 2017.

[8] D'az, A. Inside martyr complex for first-gen. https://angelddaz.substack.com/p/inside-martyr-complex-for-first-gen, December 2019.

[9] A. Gelman. Bit by bit: Social research in the digital age. http://andrewgelman.com/2018/04/14/36778/, April 2018.

[10] R. Gotian. Networking for introvered scientists. *Nature Career*, April 2019. https://www.nature.com/articles/d41586-019-01296-2.

[11] S. Huang. Finding data collection bugs: A data quality mystery. https://stat198-spring18.github.io/blog/2018/04/18/data-quality-mystery, April 2018.

[12] Icahn School of Medicine at Mount Sinai. Diversity in STEM: National organizations. https://libguides.mssm.edu/diversity/orgs.

[13] T. Jann. Understanding through documenting: The importance of seeing how your code fits into the bigger picture and how documentation can help facilitate that. https://stat198-spring18.github.io/blog/2018/04/03/understanding-through-documenting, 2018.

[14] Johns Hopkins Faculty Development. WAGs. Johns Hopkins Medicine. https://www.hopkinsmedicine.org/fac_development/career_path/wags.html.

[15] K. Mangan. Social networks for academics proliferate, despite some doubts. *The Chronicle of Higher Education*, April 2012. https://www.chronicle.com/article/Social-Networks-for-Academics/131726.

[16] A. Martinez. Science students learn to use social media to communicate research. *The Chronicle of Higher Education*, July 2016. https://www.chronicle.com/article/Science-Students-Learn-to-Use/237158.

[17] A. McNamara. Wikipedia in the classroom: Gender, argh. http://www.amelia.mn/blog/teaching/2018/03/20/Wikipedia-in-the-classroom.html, March 2018.

[18] E. Pain. How to network effectively. *Science*, October 2015. https://www.sciencemag.org/careers/2015/10/how-network-effectively.

[19] R. D. Peng. Thoughts on David Donoho's "Fifty Years of Data Science". https://simplystatistics.org/2017/12/20/thoughts-on-david-donoho-s-fifty-years-of-data-science/, December 2017.

[20] K. A. Rockquemore. Shut up and write. *Inside Higher Ed*, June 2010. https://www.insidehighered.com/advice/2010/06/14/shut-and-write.

[21] rOpenSci. Code of conduct. https://ropensci.org/code-of-conduct/, January 2020.

[22] M. Salam. Most wikipedia profiles are of men. This scientist is changing that. *The New York Times*, July 2019. http://www.nytimes.com/packages/pdf/politics/pollingstandards.pdf.

[23] SciPy Community. NumPy Code of Conduct. https://numpy.org/devdocs/dev/conduct/code_of_conduct.html, March 2020.

[24] Software Sustainability Institute. Giving a good lightning talk. https://software.ac.uk/home/cw11/giving-good-lightning-talk, 2011.

[25] Stack Overflow. How to create a minimal, reproducible example. https://stackoverflow.com/help/minimal-reproducible-example.

[26] STAT 198. About STAT 198: Blogging for data science. https://stat198-spring18.github.io/about/.

[27] S. Stoudt. Communicating with Data tweets. https://twitter.com/i/events/1230569061207031808, 2017.

[28] SusunW. Why I write about women on Wikipedia. https://blog.wikimedia.org/2018/05/03/why-i-women-wikipedia/, May 2018.

[29] Wikipedia. Template: Biography. https://en.wikipedia.org/wiki/Template:Biography, December 2019.

[30] Wikipedia. List of African-American inventors and scientists . https://en.wikipedia.org/wiki/List_of_African-American_inventors_and_scientists, February 2020.

[31] Wikipedia. List of female scientists before the 20th century. https://en.wikipedia.org/wiki/List_of_female_scientists_before_the_20th_century, January 2020.

[32] Wikipedia. List of female scientists in the 20th century. https://en.wikipedia.org/wiki/List_of_female_scientists_in_the_20th_century, January 2020.

[33] Wikipedia. List of female scientists in the 21st century. https://en.wikipedia.org/wiki/List_of_female_scientists_in_the_21st_century, February 2020.

[34] Wikipedia. Wikipedia: Biographies of living persons. https://en.wikipedia.org/wiki/Wikipedia:Biographies_of_living_persons, February 2020.

[35] Wikipedia. Wikipedia:Contributing to Wikipedia. https://en.wikipedia.org/wiki/Wikipedia:Contributing_to_Wikipedia, February 2020.

11.7 Activities

11.7.1 Interview and Write a Bio

To help make your writing presence more personal, consider adding a bio to your promotional materials. It can be hard to write your own bio so consider enlisting a peer to interview you and write a draft (and vice versa). You can then use their draft as a starting point; that is, build off of their draft to create your final bio.

When you interview your peer, go beyond the typical content included on a resumé. What are their hobbies? What personal communities do they belong to? What got them interested in their current field/career? What makes them unique? Are there any short, yet relevant, anecdotes that would help show their personality?

We have included two sample bios below.

B is a fourth year math major at UC Berkeley, but is focused on statistics after being inspired by S! He loves probability, really digs distributions, and tends to spend his free time playing video games like League of Legends and Hearthstone. He is a good person because he prefers crunchy peanut butter over smooth peanut butter. As an Oakland native of Cantonese descent, he wants everyone to know that the most useful phrase in Cantonese is "Gong hei fot choi!" because it can get you money during Chinese New Year. He works as a math tutor and hopes to spread the idea that math is cool to kids. He hopes to educate others as well as himself through these blog posts!

K is a data enthusiast and applied mathematics student at UC Berkeley in her junior year. She is interested in sharing succinct and insightful ideas that relate to data for fellow data wranglers and data science learners in an interesting and accessible manner. K has worked with local businesses and is a Data For Good competition participant who believes there is big potential in the details of analysis, and is eager to add her thoughts to the data discussion.

11.7.2 Branching Out: Brainstorm Ideas Beyond Current Work

You may want to practice writing without having to talk about your main work all of the time. Get creative when brainstorming other ideas.

- Write about a fun data analysis. If you come across an interesting dataset, you could play around with it and talk about what you find in a blog post. This does not have to be a formal or fancy data analysis (although it should be statistically sound). For the sake of brevity, the analysis will also not be comprehensive. As you explore, see what detail interests you the most and focus on that aspect.

For example, a student of ours, Huang [11], discusses a data mystery he ultimately solved.

- Write about a personal experience. Consider your experiences such as teaching, summer opportunities, or mentorship. What did you learn from each experience? Do you have advice for others in a similar situation?

  For example, a student of ours, Jann [13], recounts a personal story that taught her about her own code-writing habits.

- Synthesize something you read or heard. By sharing what you took away from a series of papers you read, a class you took, a talk, event, or conference you attended, you can connect with others interested in the topic.

  For example in a blog post, Peng [19] summarizes some commentary by Donoho and points to aspects of the commentary that resonate with him.

- Start a discussion about something you care about. Creating a forum for discussion can be another way to expand your network. A blog post aimed at such a discussion can provide a venue for debate and allow you to dig into the nuances of a particular topic.

  For example in a blog post, Gelman [9] discusses a new textbook, explaining what they like and what could be improved.

11.7.3 Wikipedia Entry for an Underrepresented Statistician

Both gender and racial bias has been a concern on Wikipedia. Biases affect who has biographical entries written about them. Your contribution to Wikipedia does not have to be technical. You can also contribute to biographical entries for people in your field and help mitigate representation bias at the same time. Writing a biographical entry will involve some research as you will not be able to rely on your own knowledge of the subject to get started. However, Wikipedia provides a template for a biography entry [29], making it easier to get started. Wikipedia also provides particular guidelines for people who are still living [34].

For inspiration, Wikipedia provides lists of women in science [31–33] and a list of African-American inventors and scientists [30].

To read about others' experiences contributing to biographical Wikipedia entries, read blog posts from McNamara [17], SusunW [28], and *The New York Times* article about Wade [22].

11.7.4 Getting Started on Social Media

Create yourself a presence on a social media platform of your choice. It takes some time to curate your account and build your positing confidence, but making some concrete goals for a week can get you started.

- Setup. If you do not have an account on a social media platform like Twitter, create one.

- Support your network. Find a peer to be your social media buddy. Keep track of their posts daily.

- Posting. Post at least once a day for a week. Posts related to supporting your peer do not count towards this count. Consider promoting your own work/writing. Consider writing your biggest take-away so far from this book and, if you are on Twitter, tag us: @DebAtStat and @sastoudt.

- Follow others. Keep track of at least 10 science-communications-related users. Note what they are posting about this week and write a summary. What is the balance between "science talk" and personal content? Do you see a network effect (i.e., do these people interact with others you are tracking)?

- Reflect on benefits. Did any opportunities (papers to read, things to apply for, chances for collaboration, etc.) come to you through this process?

11.7.5 Post About What You Are Learning

As you work your way through *Communicating with Data: The Art of Writing for Data Science*, try synthesizing what you learned and posting it on social media. This can include things that still confuse or things you would have liked to see that are missing. If you are using Twitter, tag us if you wish; we would love to read your thoughts (@DebAtStat and @sastoudt).

When we taught a course that inspired this book, Stoudt tweeted a summary after every class [27].

12 Building a Portfolio

Comfort from identifying as a writer and embracing this identity comes from practicing your craft and gaining confidence in your skills. For this, it can be helpful to build a portfolio of your written work that you can reference. You can add your work from each chapter's activities to your portfolio. In addition, this chapter includes exercises and activities to help you further practice your writing skills and build writing samples. We hope that you can leverage this body of work into a myriad of opportunities that will help you grow as a scientist.

We encourage you to work on the portfolio pieces that most interest you and to continuously refine your drafts by going through the revision process discussed in Chapter 10 and by soliciting feedback from your peers. We provide general length-guidelines for each piece, but you should feel free to adapt these. Each exercise appears on its own page, and the lengthier ones are marked with a +.

Communicating with Data: The Art of Writing for Data Science. Deborah Nolan and Sara Stoudt, Oxford University Press (2021). © Deborah Nolan and Sara Stoudt.
DOI: 10.1093/oso/9780198862741.003.0012

A Question Generation and the Ideal Data

Overview: For this exercise, you are to come up with a scientific/social question that interests you that you do not know the answer to and discuss what data you would need to collect in order to answer your question.

Prompts: Begin by stating your question and explaining why this question interests you. Then, describe the data needed to address your question by answering the following prompts.

- What would the setup be for the data collection (type of study)?

- What is the setting of the collection (where and when)?

- What/who is an observation?

- Who would collect the data?

- What information would you need to measure or collect?

- Are there background, historical, indirect, or control data that may be useful? What information would you need in order to find this data?

- Give a rough estimate of how many observations you would need for a sense of scale: tens, hundreds, thousands?

- What could go wrong in the data collection that would lead to problems in answering your question?

- How well would the data from your study generalize to other similar situations? Think of a related question that can and cannot be answered with your data.

Length: 400–600 words.

B Analyst Mindset⁺

Overview: For this activity, you are to watch episodes Roger Peng's "Analyst Mindset: Live Annotated Data Analysis" [26] and comment on his data analysis process. This activity has three parts that can be done independently. The last part requires more-advanced statistical knowledge.

Part One: Watch Episode 2, called "Wildfire Smoke Composition" and use the following prompts to summarize that episode.
Prompts:

- Does Peng do anything that you haven't seen before? Does he use any tricks that stick with you?

- Discuss Peng's use of trial and error.

- Discuss Peng's use of exploratory data analysis (EDA) to prompt questions and how he answers these questions.

- Would you have done anything differently?

- Do you have any habits/strategies in common with Peng?

- Are there any questions that Peng did not answer that could be answered given the data?

Length: 400–600 words.

Part Two: Watch all three episodes ("Analyzing Lengths of Not So Standard Deviations Episodes", "Wildfire Smoke Composition", and "Environmental Clinical Trial"). Use the following prompts to compare and contrast them.
Prompts:

- Are there common strategies/techniques across the videos?

- Which do you find most compelling or useful and why?

Length: 200–400 words.

Part Three: The webpage for Episode 3, "Environmental Clinical Trial", includes the dataset from Peng [27] and the paper by Matsui et al. [21]. Use these data to carry out your own data analysis and write a blog post. Follow the instructions below.

Instructions:

- Watch Episode 3.

- Read the paper that goes along with the episode.

- Episode 3 uses generalized estimating equations. To help familiarize yourself with this method, read an overview of the method by Hong and Ottoboni [16].

- Familiarize yourself with the data.

  - Try to reproduce Peng's results and plots.

  - Dig into different questions that were not addressed in the video.

- Prepare your findings as a blog post.

  - Include visuals.

  - Include reproducible and well-commented code.

Length: 500–1000 words.

C Book Report[+]

Overview: For this portfolio piece, read a pop-statistics or pop-science book and write a book report on it. A list of possible books to read is provided below.

Prompts: Address the following prompts in your report, in any order. Prepare your report as an essay in paragraph form and not simply as a bulleted list of answers.

- Give a brief summary of the content. The rest of the book report should focus on the writing itself rather than the content.

- Read the preface. What is the author's motivation for writing this book?

- Conduct a brief internet search about the author. What else have they written? What is their background? How does this impact their point of view?

- Who is the intended audience? How successful is the book in being accessible to its intended audience?

- Does the author have to sacrifice preciseness to maintain accessibility at certain points in the book?

- Does the author use jargon without defining it? Was there any word choice that you found confusing?

- Discuss the author's writing style, organization, and structure.

  - How does the author hook you in the beginning?

  - How is the pace of the book? Are you engaged the whole way through, or are there lulls?

  - How does the author tie things up at the end?

  - Are chapters more or less stand-alone or are there explicit transitions between them?

- Describe the major takeaways/themes.

  - What stuck with you or surprised you? How did the author craft these messages?

  - Do you disagree with anything the author says? Why?

Length: 1000–1500 words.

Book List:

Escape from the Ivory Tower: A Guide to Making Your Science Matter by Nancy Baron

The Sabermetric Revolution: Assessing the Growth of Analytics in Baseball by Benjamin Baumer and Andrew Zimbalist

Computer Age Statistical Inference: Algorithms, Evidence, and Data Science (Institute of Mathematical Statistics Monographs) by Bradley Efron and Trevor Hastie

How to Lie with Statistics by Darrell Huff and Irving Geis

The Cartoon Guide to Statistics by Larry Gonick and Woollcott Smith

The Elements of Data Analytic Style by Jeff Leek

Don't Be Such a Scientist: Talking Substance in an Age of Style by Randy Olson

Houston, We Have a Narrative: Why Science Needs Story by Randy Olson

Innumeracy: Mathematical Illiteracy and its Consequences by John Allen Paulos

A Mathematician Reads the Newspaper by John Allen Paulos

Dear Data by Giorgia Lupi and Stefanie Posavec

The Theory That Would Not Die: How Bayes' Rule Cracked the Enigma Code, Hunted Down Russian Submarines, and Emerged Triumphant from Two Centuries of Controversy by Sharon Bertsch McGrayne

Weapons of Math Destruction: How Big Data Increases Inequality and Threatens Democracy by Cathy O'Neil

Statistics Done Wrong: The Woefully Complete Guide by Alex Reinhart

Dataclysm: Who We Are (When We Think No One's Looking) by Christian Rudder

Errors, Blunders, and Lies: How to Tell the Difference (ASA–CRC Series on Statistical Reasoning in Science and Society) by David Salsburg

The Lady Tasting Tea: How Statistics Revolutionized Science in the Twentieth Century by David Salsburg

The Signal and the Noise: Why So Many Predictions Fail—But Some Don't by Nate Silver

Everybody Lies: Big Data, New Data, and What the Internet Can Tell Us About Who We Really Are by Seth Stephens-Davidowitz

The Seven Pillars of Statistical Wisdom by Stephen M. Stigler

The Black Swan: The Impact of the Highly Improbable by Nassim Nicholas Taleb

The Visual Display of Quantitative Information by Edward Tufte

Naked Statistics: Stripping the Dread from the Data by Charles Wheelan

D Article Critique

Overview: For this portfolio piece, you are to read and critique a science article/report and focus on the organization, clarity, accessibility, and style rather than the content per se. A short list of possible articles to read is provided.

Prompts: The prompts for critiquing the article are organized into three categories—entire article, individual sentences, and specific words. Address all three categories in your analysis.

At the article level:

- How would you have organized the content? Make a brief outline.

- Is anything missing (details, more thorough explanation)?

- Could anything be cut without sacrificing quality?

- Select a paragraph that you think would be improved by reorganization, and rewrite it.

- If you had to give the paper a new title, what would it be?

At the sentence level:

- Are any transitions weak? How would you strengthen them?

- Are there any run-on sentences? How would you break them up?

- Pick a few sentences and rewrite them, noting what you changed and why.

At the word level:

- Are there any words that are over-complicated? Is there a simpler way of saying things?

- Choose five keywords to summarize the content.

Length: 400–600 words.

Article List:

"Statistical fraud detection: A review" by Richard J. Bolton and David J. Hand [6]

"County-level analysis of the impact of temperature and population increases on California wildfire data" by M. Baltar, J. E. Keeley, and F. P. Schoenberg [5]

"Glyph-maps for visually exploring temporal patterns in climate data and models" by Hadley Wickham, Heike Hofmann, Charlotte Wickham, and Dianne Cook [34]

"Estimating the prevalence of hard-to-reach populations: The illustration of mark-recapture methods in the study of homelessness" by Ian Shaw [29]

"Building an NCAA men's basketball predictive model and quantifying its success" by Michael J. Lopez and Gregory J. Matthews [19]

E Scope of a Journal⁺

Overview: For this activity, you are to compare and contrast a journal from each of two different types of journals. Base your comparison on reading each journal's published "aims and scope" and abstracts of the articles in the latest issue.

Prompts: Select two journals that you are interested in reading or publishing in. As mentioned above, these journals should be from different categories (see below for sample categories and journals). Read the abstracts from the most recent issue and the aims and scope of each journal. Use the following questions to describe the scope of the journals.

- Who is the intended audience for each journal? Are the abstracts accessible to that audience?

- For each abstract, what is the main focus: data, methods, computation, theory, something else?

- Do the articles in the latest issue seem to have a common theme (what is it?) or do they cover a wide range of topics?

- In your opinion, do the abstracts match up with the defined scope of the journal?

- Change part of one of the abstracts to either fit the scope better or be more accessible to a wider audience.

- Is there a common structure to the abstracts themselves? Are these structures similar between the two journals?

Length: 400–600 words.

Journal Categories:

Theoretical statistics: *Journal of the American Statistical Association, Annals of Statistics, Biometrika*

Applied statistics: *Environmetrics, The American Statistician, The Annals of Applied Statistics, Statistical Science*

Computational statistics: *Journal of Statistical Software, The R Journal, Journal of Computational and Graphical Statistics*

General science: *Proceedings of the National Academy of Sciences of the United States of America, Nature, Science, PLOS ONE*

Life science: *Cell, eLife, Ecology Letters, Trends in Ecology & Evolution*

Medicine and Public Health: *Journal of the American Medical Association, The Lancet, American Journal of Public Health, Annual Review of Nutrition*

Physical science: *Astronomy and Astrophysics, Chemical Reviews, Reviews of Modern Physics, Bulletin of the American Meteorological Society*

F xkcd Common Words

Overview: This tongue-in-cheek portfolio piece has you write an abstract for an article/report without using any technical words. The xkcd "Simple Writer" app (https: //xkcd.com/simplewriter/) checks your words to see if they are among the 1000 most commonly used. (See for example: https://xkcd.com/1133/.)

Prompt: Select a scientific abstract to translate into common, nontechnical words. (You may instead choose to summarize a report that you have written.) Rewrite the abstract using only those words that pass the "xkcd" test. (If you absolutely cannot avoid using a word that is not a common word, underline that word and write a few sentences about your attempts to get around using the word.)

Length: 200–300 words.

Here is an example based on "Student evaluations of teaching (mostly) do not measure teaching effectiveness" by Anne Boring, Kellie Ottoboni, and Philip B. Stark [7]. The abstract for this article appears below.

Student evaluations of teaching (SET) are widely used in academic personnel decisions as a measure of teaching effectiveness. We show:

- SET are biased against female instructors by an amount that is large and statistically significant.
- The bias affects how students rate even putatively objective aspects of teaching, such as how promptly assignments are graded.
- The bias varies by discipline and by student gender, among other things.
- It is not possible to adjust for the bias, because it depends on so many factors.
- SET are more sensitive to students' gender bias and grade expectations than they are to teaching effectiveness.
- Gender biases can be large enough to cause more effective instructors to get lower SET than less effective instructors.

These findings are based on nonparametric statistical tests applied to two datasets: 23,001 SET of 379 instructors by 4,423 students in six mandatory first-year courses in a five-year natural experiment at a French university, and 43 SET for four sections of an online course in a randomized, controlled, blind experiment at a US university.

Our alternative abstract reads as follows:

Students fill out forms about their teachers to say how good or bad they think they are. These forms get used by almost all schools to decide if the teachers get to keep their jobs. We show:

- The forms aren't the same for women and men teachers. Men get much higher (better) numbers than women even when they are just as good as each other.

- When the students say how quick the teacher is to give back their work, they say men are better even though in fact the men and women are just as fast.
- Girl and boy students give different numbers for how good the same teacher is.
- The men do better on the forms than the women teachers for lots of groups of students, not just boys and girls.
- How much better or worse a student says a women teacher is than a man teacher can be explained more by what the student thinks they will get for their class work than on how good the teacher really is.
- A good women teacher can even get worse numbers than a bad man teacher.

We found all of this out from studying two sets of students. One set has 4,423 students fill in forms for 379 teachers. This set has 23,001 forms for 6 courses over 5 years. The other set has 43 students across 4 classes. We use chance to tell students which class to go to. In this small set, the students do not know if their teacher is really a man or a woman. They just know the teacher's name. They never see the teacher, and the same teacher uses both girl and boy names.

G Listening Tour[*]

Overview: For this activity, attend a scientific seminar outside of your field and write a big-picture narrative of the topic and the data the speaker has analyzed.

Prompts: Choose a seminar that interests you and where the speaker uses data to answer a question. The seminar can be outside of your comfort zone content-wise; that is, you don't need to understand the details of the talk. Write an overview of the seminar in nontechnical language. Use the following prompts to organize your report.

- Data.

    - What kind of data does the speaker work with?

    - Has the speaker collected the data? How?

- Question.

    - What is the problem the scientist is grappling with?

    - Why is the problem important?

- Solution.

    - How is the speaker addressing the problem?

    - Are there any simplifying assumptions made?

    - What still needs to be done?

- Presentation.

    - Was the seminar presented in a way that you could understand? Which part(s) were comprehensible?

    - How was the pace and organization of the talk?

    - Were there any figures or images that were particularly effective?

Length: 400–600 words.

H Write an Abstract

Overview: For this activity, you are to read a scientific article, write an abstract for it, and compare your abstract to the author's.

Instructions:

- Choose an article of interest. A short list of possible articles to read appears at the end of this activity.

- Read the article, but DO NOT read the abstract the author(s) wrote. You may want to print the paper and cover up or remove the abstract.

- Write an abstract for the article.

- *After* you have written your abstract, read the real abstract for the article.

- Write a paragraph that compares and contrasts your abstract with the real one. What did you do similarly/differently?

Length: 150 words for the abstract, 200–300 words for the comparison.

Article List:

"The perils of cherry picking low frequency events in large sample surveys" by Stephen Ansolabehere, Samantha Luks, and Brian F. Schaffner [2]

"Anthropogenic warming has increased drought risk in California" by Noah S. Diffenbaugh, Daniel L. Swain, and Danielle Touma [12]

"Climate and infectious disease: Use of remote sensing for detection of *Vibrio cholerae* by indirect measurement" by Brad Lobitz, Louisa Beck, Anwar Huq, Byron Wood, George Fuchs, A. S. G. Faugue, and Rita Colwell [18]

"The butterfly did it: The aberrant vote for Buchanan in Palm Beach County, Florida" by Jonathan N. Wand, Ken Shotts, Jasjeet S. Sekhon, Walter R. Mebane Jr., Michael C. Herron, and Henry E. Brady [33]

"Killer whale presence drives bowhead whale selection for sea ice in Arctic seascapes of fear" by Cory J. D. Matthews, Greg A. Breed, Bernard Le Blanc, and Steven H. Ferguson [22]

"Exercise effects on bed rest-induced brain changes" by Vincent Koppelmans, Jessica M. Scott, Meghan E. Downs, Kaitlin E. Cassady, Peng Yuan, Ofer Pasternak, Scott J. Wood, Yiri E. De Dios, Nichole E. Gadd, Igor Kofman, Roy Riascos, Patricia A. Reuter-Lorenz, Jacob J. Bloomberg, Ajitkumar P. Mulavara, Lori L. Ploutz-Snyder, and Rachael D. Seidler [17]

"Predicting the number of defects in a new software version" by Ebubeougu Amarachukwu Felix and Sai Peck Lee [13]

"Modeling environmental variability and network formation among pastoral nomadic households: Implications for the rise of the Mongol Empire" by Daniel R. Schultz and Andrew Costopoulos [30]

I Dear Data⁺

Overview: For this exercise, you track your actions for a week and create a visualization of them.

Instructions:

- Look through the visualizations in *Dear Data* by Lupi and Posavec [20]. The visualizations in Shaffer and Kriebel [28] and Avirgan and LaRue from *Five Thirty Eight* [4] may also provide inspiration.

- Choose something about yourself to keep track of and collect data for one week.

- Draw a colorful visualization of your data.

- Write a caption to explain your visualization, including the data, use of symbols, and interesting features.

Length: one hand-drawn visualization and 50–100 word caption.

J Effective Examples and Analogies

Overview: For this exercise, you are to discuss an example or analogy that has particularly stuck with you as you learned technical material.

Prompts: Write a brief description of an analogy or example that you have found to be effective in learning something technical. Then answer the following questions.

- How did you come across the example?

- Does it help you remember something or did it help you understand something?

- What makes it memorable/understandable?

- Have you passed this example along to anyone? Did it help them?

Length: 150–250 words.

K Revise and Resubmit*

Overview: For this activity, you are to formalize the feedback you get from two reviewers and prepare a "revise and resubmit" response where you state what you changed, where and why.

Prompts: When we submit a paper to a journal, we often receive a request to "revise and resubmit". Along with submitting a revised manuscript, we are also expected to provide a document that describes in detail how we addressed (or chose not to address) each of the comments from the editor(s) and reviewer(s).

If you have not already done so, ask two people (e.g., a peer and an instructor) to read and review a draft of an article/blog post/press release that you have written.

Follow the following guidelines to prepare a revise-and-resubmit response to the reviews.

- Make a list of all of the feedback you received from your reviewers. This list can include specific things such as typos, missing commas, etc., as well as more general feedback such as the need for stronger transitions or a different methodology for a certain analysis.

- Revise your paper to incorporate the requested changes. You may elect not to incorporate some of the suggestions. If you received contradictory or overlapping suggestions, it is up to you to address the contradiction and explain why you made the revision you did.

- Write an introductory paragraph that summarizes the big-picture changes you made to your paper.

- Then use the list of the feedback that you compiled as a starting point to discuss how you addressed each piece of feedback. This part of the response can remain in list form.

- Be sure to note the place in the document where each change was made, e.g., include page number, and line/paragraph number where the revision occurs.

- For simple typos, you can simply state that these have been fixed.

- Justify your decision to not incorporate a reviewer's suggestion and explain why you chose to revise differently than what was requested.

Length: As long as needed to address each piece of feedback (typically, 1 or 2 single-spaced pages).

L Science in the News[+]

Overview: For this portfolio piece, you are to track science stories from a particular news source for a week and comment on what they choose to report, how they present it, etc.

Prompts: Choose a news source and follow it for a week. A list of possible sources appears at the end of this exercise. Take notes of the topics, authors, and findings using the prompts below as a guide. Write a summary of your findings.

- What topics are covered? Do the topics follow a common theme?

- Who are the writers and what are their backgrounds?

- Is the way the writers cover certain topics different than for other topics?

- Were the results overstated?

- What do you notice about the vocabulary choice?

- What were the writing strategies used to draw you in and keep you interested?

- After reading the articles do you feel like you understand the content? Are there questions you have that are left unanswered?

Length: 400–600 words.

Possible News Sources:

The Science & technology Section in *The Economist*

The Science Section in *The New York Times*

The Health and Science Sections in *The Washington Post*

Scientific American

The Science & Environment Section in the *BBC News*

Transcripts of On Science podcasts on *National Public Radio*

Transcripts of *Radiolab* podcasts on *National Public Radio*

Science News

M Interactive Visualizations*

Overview: For this activity, you are to play around with an interactive visualization and note what you find beyond what is written in the associated news story or blog post.

Prompts: Choose an interactive visualization that interests you from a news source or blog. (Suggestions appear at the end of this activity.) Use the following prompts to structure your interactions and investigations of the visualization and to write about your findings.

- What was your strategy for looking around in the visualized data?

- What questions motivated your perusing? Were you able to answer them?

- Was the visualization set up in a way that made it easy to look for insights?

- How would you improve the visualization?

- What other datasets would benefit from this kind of visualization?

- What did you find and what story can you tell given your discoveries?

Length: 250–350 words.

Possible Interactive Visualizations:

"How the Internet* talks" by Randy Olson and Ritchie King (*Five Thirty Eight* [25])

"Science isn't broken" by Christie Aschwanden and Ritchie King (*Five Thirty Eight* [3])

"Music Timeline" (Google [15])

"Gender Pay Gap" by David McCandless et al. (information is beautiful [23])

"A day in the life of Americans" by Nathan Yau (*Flowing Data [35]*)

*"A day in the life: Work and home" by Nathan Yau (*Flowing Data [36])

"You draw it: What got better or worse during Obama's presidency" by Larry Buchanan, Haeyoun Park, and Adam Pearce (*The New York Times* [8])

"The voting habits of Americans like you" by Nate Cohn and Amanda Cox (*The New York Times* [10])

"Where the poor live longer: How your area compares" by Gregor Aisch et al. (*The New York Times* [1])

"Fractions of a second: An Olympic musical" by Amanda Cox (*The New York Times* [11])

"How different groups spend their day" by Shan Carter et al. (*The New York Times* [9])

N Five Thirty Eight Investigation[*]

Overview: In this portfolio piece, you are to read a *Five Thirty Eight* article that has a dataset supplied on their GitHub page, explore the data, and write up what you find.

Prompts: Choose an article from *Five Thirty Eight* [14] that interests you. Be sure to select an article where the data are available. Use the following prompts to structure your investigations.

- Try to reproduce a figure or a statistical result in the article.

- If you are unable to reproduce it or find a discrepancy, show your code to someone else and get help figuring out the problem.

- Explore the data further. Find something else of interest that adds to the story or starts a new story.

Length: 300–500 words, code used to reproduce the result from the article and to explore the data further.

O Turning a Portfolio Piece into a Blog Post⁺

Overview: For this activity, you are to revisit one of your portfolio pieces and create a blog post from it.

Prompts: Choose one of your completed portfolio pieces to turn into a blog. Some pieces that make good material for blogging are Portfolio Pieces A, C, G, I, and N. After you write your blog post, reflect on the process using the prompts below.

- Was it easier/harder/about the same to write the blog post than it was to write the original piece?

- What was your strategy for making your original piece more accessible?

- Does the structure/ordering change?

- What did you choose to emphasize/de-emphasize?

Length: 250 words for the prompt; 300–600 words for the blog post.

P Blog Narrative Arc

Overview: For this activity, you pick a blog that has an established record of posts and write about the themes and the narrative of the blog as a whole.

Prompts: Choose a science-related blog that interests you and that you read regularly. (Below is a list of possible blogs, if you are not a regular reader.) Skim a few dozen posts over an extended period of time, and follow the prompts below to summarize the blog.

- Is there a common theme to the posts (what is it?) or do the topics vary?

- Are there natural groupings of posts or is each stand-alone?

- Look up the author. How does the author's background shape the perspective and topics they choose to blog about?

- Does the author reply to feedback on their posts?

- Select five of the blog posts and write three keywords for each.

Length: 300–500 words.

List of Science-Related Blogs:

https://astrobites.org/
https://bair.berkeley.edu/blog/
https://blogs.scientificamerican.com/
https://chembites.org/
https://ecologyforthemasses.com/
https://envirobites.org/
https://statmodeling.stat.columbia.edu/

Q Blog Critique

Overview: For this exercise, you are to critique, and compare and contrast two blog posts.

Prompts: Choose two blog posts of interest to you; they can be from the same blog or from two separate blogs. (Portfolio Piece P provides a list of science-related blogs, if needed.) Use the following prompts to compare and contrast the posts.

- Do the two posts address similar topics?

- Which post is more accessible and/or compelling? What makes it so? Is it the topic, the style, the examples, the visuals, or something else?

- Compare and contrast their styles. Is one post more formal than the other? Is one more technical than the other? If they are from the same blog, does the style change between the two posts?

- Compare and contrast the posts' structure and organization. Where is the punch-line in each of the posts?

- Find a weak piece in each blog post and rewrite it.

- Pick one of the posts and write a follow-up comment. Is there a part you disagree with? Is there something that remains unclear? Is there something that is particularly clear or interesting?

Length: 400–600 words.

R Pictoral Key to a Blog Post*

Overview: For this portfolio piece, you are to write a blog post about a "one-minute revelation" plot.

Prompts: Blog posts can benefit from the use of one or two key pictures that help emphasize the findings as well as grab a reader's attention. As an extension of Section 4.8.2, write a blog post about your one-minute-revelation graph. Use the following prompts to guide you.

- Fully describe the visualization in the post.

- Challenge yourself to write a post that goes beyond a summary of the plot. What angle adds more to the narrative?

- Figure captions are less common in blog-post figures, but use your judgment.

Length: 300–600 words.

S Critique Wiki Entries[+]

Overview: For this exercise, you are to discuss and critique Wikipedia entries for topics related to data analysis.

Prompts: Find two Wikipedia pages that you think are successful and two that you think are unsuccessful. Use the following prompts to write your critique.

- Successful pages.

  - What makes the page successful? Discuss style, pace, word choice, clarity, and quality of citations.

  - Is there a figure, image, or diagram that is particularly effective? Describe the value it adds to the page.

  - Do you think part of the success comes from the topic itself being "easy" or is it successful because it synthesizes a complicated topic well?

- Unsuccessful pages.

  - Is the page incomplete? What is missing?

  - Is the page confusing? How would you rewrite a part that you find particularly confusing?

  - Is the page vague or unclear? How would you rewrite a part that you find vague or unclear?

  - Is the page missing citations? Find a potential reference.

  - Would the page benefit from a figure, image, or diagram? Describe what would be helpful.

Length: 300–500 words.

T Tidy Tuesday Visualization⁺

Overview: Each week the *R for Data Science* online learning community posts a dataset and a published plot made from the dataset. The goal of Tidy Tuesday is to improve the graphic or display some other aspect of the dataset and share your results (code and figure) via GitHub and Twitter. For this portfolio assignment, you are to participate in a Tidy Tuesday (past weeks are fine). Note you can use Python to analyze the data if you prefer.

Read more about Tidy Tuesday and find the weekly posts on the GitHub repository [24].

Prompts: Create a new statistical graph of the data. As mentioned above, this plot should offer a new insight or improve upon the graphic supplied in the post. Answer the following prompts in the summary of your work.

- What could be improved in the original figure?

- How did you address those weaknesses in your own approach?

- What problems did you face as you tried to manipulate or display the data?

- What is the take-away from your plot?

- If you had more time, what would you explore next or perfect in terms of the visualization?

Length: a paragraph summary of your approach to the week's figure, plus the code to create your figure, and the figure.

U Diary Entry to Blog Post⁺

Overview: For this portfolio piece, take one of the ideas from your brainstorming diary entries (see Section 7.8.1) and turn it into a blog post.

Instructions: To help you select a topic for your blog post, review your brainstorming diary as follows:

- Share your diary entries with a peer. Get their feedback about what sounds interesting and see if they have any follow-up ideas.

- Decide which topic seems the most interesting yet actionable. Avoid picking too broad of a topic or a question that is not answerable given your access to data, resources, etc.

- If your idea requires data analysis, research, etc., write down a specific goal for the post first before starting to explore. This will help you stay focused. You want to spend most of your energy on writing, not doing fancy analysis (although it should be statistically sound).

As you write, if you have a new, yet related, idea, feel free to change the focus of the blog post. However, write down your new goal and make sure you remain focused on it. You do not want a blog post that starts with one goal in mind and then goes off on a tangent.

Length: 300–600 words.

V Q&A in an Online-Learning Community

Overview: For this activity, you are to engage in an online learning community, such as Stack Overflow [32] or Stack Exchange [31]. Online communities help bring people together for a common purpose. For communities whose primary goal is education, they provide a forum for users to ask questions and contribute their own knowledge by answering others' questions. On Stack Overflow, users can get coding help for a variety of programming languages, and on Stack Exchange, questions beyond code are asked and answered.

Instructions: Participate in Stack Overflow or Stack Exchange by answering and/or asking five questions. Use any of the following approaches.

- Pick a statistical or coding concept that you feel comfortable with and respond to any unanswered questions you know the answer to.

- Refine an answer to a question that you think can be improved.

- Pick a statistical or coding concept you want to learn more about and propose a question you have. Make sure it hasn't already been asked and answered.

Length: at least five posts (not including follow-up).

W References

[1] G. Aisch, Q. Bui, A. Cox, and K. Quealy. Where the poor live longer: How your area compares. *The New York Times*, April 2016. https://www.nytimes.com/interactive/2016/04/11/upshot/where-the-poor-live-longer-how-your-area-compares.html.

[2] S. Ansolabehere, S. Luks, and B. F. Schaffner. The perils of cherry picking low frequency events in large sample surveys. *Electoral Studies*, 40:409–410, 2015.

[3] C. Aschwanden and R. King. Science isn't broken. *Five Thirty Eight*, August 2015. https://fivethirtyeight.com/features/science-isnt-broken/.

[4] J. Avirgan and K. LaRue. We asked you to visualize your podcast listening and, wow, did you deliver. *Five Thirty Eight*, December 2016. https://fivethirtyeight.com/features/we-asked-you-to-visualize-your-podcast-listening-and-wow-did-you-deliver/.

[5] M. Baltar, J. E. Keeley, and F. P. Schoenberg. County-level analysis of the impact of temperature and population increases on California wildfire data. *Environmetrics*, 25(6):397–405, 2014.

[6] R. J. Bolton and D. J. Hand. Statistical fraud detection: A review. *Statistical Science*, 17(3):235–249, 2002.

[7] A. Boring, K. Ottoboni, and P. B. Stark. Student evaluations of teaching (mostly) do not measure teaching effectiveness. *ScienceOpen Research*, 2016.

[8] L. Buchanan, H. Park, and A. Pearce. You draw it: What got better or worse during Obama's presidency. *The New York Times*, January 2017. https://www.nytimes.com/interactive/2017/01/15/us/politics/you-draw-obama-legacy.html.

[9] S. Carter, A. Cox, K. Quealy, and A. Schoenfeld. How different groups spend their day. *The New York Times*, August 2009. https://archive.nytimes.com/www.nytimes.com/interactive/2009/07/31/business/20080801-metrics-graphic.html.

[10] N. Cohn and A. Cox. The voting habits of Americans like you. *The New York Times*, June 2016. https://www.nytimes.com/interactive/2016/06/10/upshot/voting-habits-turnout-partisanship.html.

[11] A. Cox. Fractions of a second: An Olympic musical. *The New York Times*, February 2010. https://archive.nytimes.com/www.nytimes.com/interactive/2010/02/26/sports/olympics/20100226-olysymphony.html.

[12] N. S. Diffenbaugh, D. L. Swain, and D. Touma. Anthropogenic warming has increased drought risk in California. *Proceedings of the National Academy of Sciences of the United States of America*, 112(13):3931–3936, 2015.

[13] E. A. Felix and S. P. Lee. Predicting the number of defects in a new software version. *PLoS ONE*, 15(3), 2020.

[14] Five Thirty Eight. data. https://github.com/fivethirtyeight/data.

[15] Google. Music timeline. https://research.google.com/bigpicture/music/.

[16] J. Hong and K. Ottoboni. Generalized estimating equations (GEE). https://rlbarter.github.io/Practical-Statistics/2017/05/10/generalized-estimating-equations-gee/, May 2017.

[17] V. Koppelmanns, J. M. Scott, M. E. Downs, K. E. Cassady, P. Yuan, O. Pasternak, S. J. Wood, Y. E. De Dios, N. E. Gadd, I. Kofman, R. Riascos, P. A. Reuter-Lorenz, J. J. Bloomberg, A. P. Mulavara, L. L. Ploutz-Snyder, and R. D. Seidler. Exercise effects on bed rest-induced brain changes. *PLoS ONE*, 13(10), 2018.

[18] B. Lobitz, L. Beck, A. Huq, B. Wood, G. Fuchs, A. S. G. Faugue, and R. Colwell. Climate and infectious disease: Use of remote sensing for detection of *Vibrio cholerae* by indirect measurement. *Proceedings of the National Academy of Sciences of the United States of*

America, 97(4):1438–1443, 2000.

[19] M. J. Lopez and G. J. Matthews. Building an NCAA men's basketball predictive model and quantifying its success. *Journal of Quantitative Analysis in Sports*, 11(1), 2015.

[20] G. Lupi and S. Posavec. *Dear Data*. http://www.dear-data.com/theproject.

[21] E. C. Matsui, M. Perzanowski, R. D. Peng, R. A. Wise, S. Balcer-Whaley, M. Newman, A. Cunningham, A. Divjan, M. E. Bollinger, S. Zhai, G. Chew, R. L. Miller, and W. Phipatanakul. Effect of an integrated pest management intervention on asthma symptoms among mouse-sensitized children and adolescents with asthma: A randomized clinical trial. *Journal of the American Medical Association*, 317(10):1027–1036, 2017.

[22] C. J. D. Matthews, G. A. Breed, B. LeBlanc, and S. H. Ferguson. Killer whale presence drives bowhead whale selection for sea ice in arctic seascapes of fear. *Proceedings of the National Academy of Sciences of the United States of America*, 117(12):6590–6598, 2020.

[23] D. McCandless, S. Starling, M. Quick, P. Thomas, and B. Bergamaschi. Gender pay gap. *information is beautiful*, April 2016. https://www.informationisbeautiful.net/visualizations/gender-pay-gap/.

[24] T. Mock. tidytuesday. *Flowing Data*, February 2020. https://github.com/rfordatascience/tidytuesday.

[25] R. Olson and R. King. How the Internet* talks. *Five Thirty Eight*, September 2017. https://projects.fivethirtyeight.com/reddit-ngram/.

[26] R. D. Peng. Analyst mindset: Live annotated data analysis. http://analystmindset.com/.

[27] R. D. Peng. Mouse allergen and asthma intervention trial. https://rdpeng.github.io/MAAIT/.

[28] J. Shaffer and A. Kriebel. *Dear Data Two*. http://www.dear-data-two.com/.

[29] I. Shaw, M. Bloor, R. Cormack, and H. Williamson. Estimating the prevalence of hard-to-reach populations: The illustration of mark-recapture methods in the study of homelessness. *Social Policy and Administration*, 30(1):69–85, 1996.

[30] D. R. Shultz and A. Costopoulos. Modeling environmental variability and network formation among pastoral nomadic households: Implications for the rise of the Mongol Empire. *PLoS ONE*, 14(10), 2019.

[31] Stack Exchange. Welcome to Stack Exchange. https://stackexchange.com/tour.

[32] Stack Overflow. How to create a minimal, reproducible example. https://stackoverflow.com/help/minimal-reproducible-example.

[33] J. N. Wand, K. Shotts, J. S. Sekhon, W. R. Mebane Jr., M. C. Herron, and H. E. Brady. The butterfly did it: The aberrant vote for Buchanan in Palm Beach County Florida. *American Political Science Review*, 95(4):793–810, 2001.

[34] H. Wickham, H. Hofmann, C. Wicham, and D. Cood. Glyph-maps for visually exploring temporal patterns in climate data and models. *Environmetrics*, 23(5):382–393, 2012.

[35] N. Yau. A day in the life of Americans. *Flowing Data*, December 2015. http://flowingdata.com/2015/12/15/a-day-in-the-life-of-americans/.

[36] N. Yau. A day in the life: Work and home. *Flowing Data*, May 2017. http://flowingdata.com/2017/05/17/american-workday/.

Index